T0211901

Architectural and Operating System Support for Virtual Memory

Synthesis Lectures on Computer Architecture

Editor
Margaret Martonosi, *Princeton University*

Founding Editor Emeritus
Mark D. Hill, *University of Wisconsin, Madison*

Synthesis Lectures on Computer Architecture publishes 50- to 100-page publications on topics pertaining to the science and art of designing, analyzing, selecting and interconnecting hardware components to create computers that meet functional, performance and cost goals. The scope will largely follow the purview of premier computer architecture conferences, such as ISCA, HPCA, MICRO, and ASPLOS.

Security Basics for Computer Architects
Ruby B. Lee
2013

The Datacenter as a Computer: An Introduction to the Design of Warehouse-Scale
Machines, Second edition
Luiz André Barroso, Jimmy Clidaras, and Urs Hölzle
2013

Shared-Memory Synchronization
Michael L. Scott
2013

Resilient Architecture Design for Voltage Variation
Vijay Janapa Reddi and Meeta Sharma Gupta
2013

Multithreading Architecture
Mario Nemirovsky and Dean M. Tullsen
2013

Performance Analysis and Tuning for General Purpose Graphics Processing Units
(GPGPU)
Hyesoon Kim, Richard Vuduc, Sara Baghsorkhi, Jee Choi, and Wen-mei Hwu
2012

Automatic Parallelization: An Overview of Fundamental Compiler Techniques
Samuel P. Midkiff
2012

Phase Change Memory: From Devices to Systems
Moinuddin K. Qureshi, Sudhanva Gurumurthi, and Bipin Rajendran
2011

Multi-Core Cache Hierarchies
Rajeev Balasubramonian, Norman P. Jouppi, and Naveen Muralimanohar
2011

A Primer on Memory Consistency and Cache Coherence
Daniel J. Sorin, Mark D. Hill, and David A. Wood
2011

Dynamic Binary Modification: Tools, Techniques, and Applications
Kim Hazelwood
2011

Architectural and Operating System Support for Virtual Memory
Abhishek Bhattacharjee and Daniel Lustig

ISBN: 978-3-031-00629-6   paperback
ISBN: 978-3-031-01757-5   ebook

DOI 10.1007/978-3-031-01757 5

A Publication in the Springer Publishers series *SYNTHESIS LECTURES ON COMPUTER ARCHITECTURE*

Lecture #42
Series Editor: Margaret Martonosi, *Princeton University*
Founding Editor Emeritus: Mark D. Hill, *University of Wisconsin, Madison*
Series ISSN
Print 1935-3235   Electronic 1935-3243

Architectural and Operating System Support for Virtual Memory

Abhishek Bhattacharjee
Rutgers University

Daniel Lustig
NVIDIA

SYNTHESIS LECTURES ON COMPUTER ARCHITECTURE #42

ABSTRACT

This book provides computer engineers, academic researchers, new graduate students, and seasoned practitioners an end-to-end overview of virtual memory. We begin with a recap of foundational concepts and discuss not only state-of-the-art virtual memory hardware and software support available today, but also emerging research trends in this space. The span of topics covers processor microarchitecture, memory systems, operating system design, and memory allocation. We show how efficient virtual memory implementations hinge on careful hardware and software cooperation, and we discuss new research directions aimed at addressing emerging problems in this space.

Virtual memory is a classic computer science abstraction and one of the pillars of the computing revolution. It has long enabled hardware flexibility, software portability, and overall better security, to name just a few of its powerful benefits. Nearly all user-level programs today take for granted that they will have been freed from the burden of physical memory management by the hardware, the operating system, device drivers, and system libraries.

However, despite its ubiquity in systems ranging from warehouse-scale datacenters to embedded Internet of Things (IoT) devices, the overheads of virtual memory are becoming a critical performance bottleneck today. Virtual memory architectures designed for individual CPUs or even individual cores are in many cases struggling to scale up and scale out to today's systems which now increasingly include exotic hardware accelerators (such as GPUs, FPGAs, or DSPs) and emerging memory technologies (such as non-volatile memory), and which run increasingly intensive workloads (such as virtualized and/or "big data" applications). As such, many of the fundamental abstractions and implementation approaches for virtual memory are being augmented, extended, or entirely rebuilt in order to ensure that virtual memory remains viable and performant in the years to come.

KEYWORDS

virtual memory, address translation, paging, swapping, main memory, disk

Contents

Preface

This book details the current state of art of software and hardware support for virtual memory (VM). We begin with a quick recap of VM basics, and then we jump ahead to more recent developments in the VM world emerging from both academia and industry in recent years. The core of this book is dedicated to surveying the highlights and conclusions from this space. We also place an emphasis on describing some of the important open problems that are likely to dominate research in the field over the coming years. We hope that readers will find this a useful guide for choosing problems to attack in their work.

Chapter 2 summarizes the basics of the VM abstraction. It describes the layout and management of a typical virtual address space, from basic memory layouts and permissions bits to shared memory and thread-local storage. Chapter 3 then provides an overview of the implementation of a typical modern paging-based VM subsystem. These chapters serve as a refresher for anyone who might be less familiar with the material. Readers may also find it helpful to review the subtleties of topics such as synonyms and homonyms. However, more experienced readers may simply choose to skip over these chapters.

The core of the book starts in Chapters 4 and 5. These chapters explore the hardware and software design spaces, respectively, for modern VM implementations. Here we explore page table layouts, TLB arrangements, page sizes, operating system locality management techniques, and memory allocation heuristics, among other things. Chapter 6 then covers VM (non-)coherence and the challenges of synchronizing highly parallel VM implementations. These chapters emphasize how the design spaces of modern VM subsystems continue to evolve in interesting new directions in order to keep up with the ever-growing working sets of today's applications.

From here, the book shifts into more forward-looking topics in the field. Chapter 7 presents some of the ways in which virtual memory is being adapted to various kinds of architectural and memory technology heterogeneity. Chapter 8 describes some of the newest research being done to improve VM system hardware, and then Chapter 9 does the same for co-designed hardware and software. At this point, we expect the reader will be able to dive into the literature well-prepared to continue their exploration into the fast-changing world of VM, and then even to help contribute to its future!

We do assume that readers already have some appropriate background knowledge. On the computer architecture side, we assume a working knowledge of fundamental concepts such as pipelining, superscalar and out-of-order scheduling, caches, and the basics of cache coherence. On the operating systems side, we assume a basic understanding of the process and thread models of execution, the kernel structures used to support these modes of execution, and the basics of memory management and file systems.

Abhishek Bhattacharjee and Daniel Lustig
September 2017

Acknowledgments

We would like to thank several people for making this manuscript possible. Eight years ago, our advisor, Margaret Martonosi, started us down this research path. We thank her for her support in pursuing our research endeavors. We also thank the many collaborators with whom we have explored various topics pertaining to virtual memory. While there are too many to name, Arka Basu, Guilherme Cox, Babak Falsafi, Gabriel Loh, Tushar Krishna, Mark Oskin, David Nellans, Binh Pham, Bharath Pichai, Geet Sethi, Jan Vesely, and Zi Yan deserve special mention for making a direct impact on the work that appeared in this book. Thank you also to Trey Cain, Derek Hower, Lisa Hsu, Aamer Jaleel, Yatin Manerkar, Michael Pellauer, and Caroline Trippel for the countless helpful discussions about virtual memory and memory system behavior in general over the years. We also thank Arka Basu, Tushar Krishna, and an anonymous reviewer for their helpful comments and suggestions to improve the quality of this book. A special thanks to Mike Morgan for his support of this book.

On a personal note, we would like to thank Shampa Sanyal for enabling our research endeavors, and we would like to thank our respective families for making this all possible in the first place.

Abhishek Bhattacharjee and Daniel Lustig
September 2017

CHAPTER 1

Introduction

Modern computer systems at all scales—datacenters, desktops, tablets, wearables, and often even embedded systems—rely on *virtual memory* (VM) to provide a clean and practical programming model to the user. As the reader is likely aware, VM is an idealized abstraction of the storage resources that are actually available on a given machine. Programs perform memory accesses using only *virtual addresses*, and the processor and operating system work together to translate those virtual addresses into *physical addresses* that specify where the data is actually physically located (Figure 1.1). The purpose of this book is to describe both the state of the art in VM design and the open research and development questions that will guide the evolution of VM in the coming years.

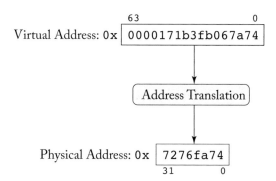

Figure 1.1: Address translation, in its most basic form.

1.1   WHY VIRTUAL MEMORY IS USED

Although we expect most readers will already have at least some background on VM basics already, we feel it is nevertheless important to begin by recapping some of the benefits of VM. These benefits are what motivate the need to continue augmenting and extending VM to be capable of supporting challenges such as architectural heterogeneity, so-called "big data," and virtualization. As such, we will need to keep them in mind as the goal for all of the new research and development being done in the field today.

The VM abstraction allows code to be written as if it has total unrestricted control of the entire available memory range, regardless of the behavior or memory usage of any other programs running concurrently in the system. This in turn allows programmers to write code that is portable with respect to changing physical resources, whether due to a dynamic change in utilization on a single machine or to a move onto a different machine with a completely different set of physical resources to begin with. In fact, user-level processes in general have no way to even determine the physical addresses that are being used behind the scenes. Without VM, programmers would have to understand the low-level complexity of the physical memory space, made up of several RAM chips, hard-disks, solid-state drives, etc., in order to write code. Every change in RAM capacity or configuration would require programmers to rewrite and recompile their code.

VM also provides protection and isolation, as it prevents buggy and/or malicious code or devices from touching the memory spaces of other running programs to which they should not have access. Without VM (or other similar abstractions [112]), there would be no memory protection, and programs would be able to overwrite and hence corrupt memory images of other programs. Security would be severely compromised as malicious programs would be able to corrupt the memory images of other programs.

Next, VM improves efficiency by allowing programs to undersubscribe (use less memory than they allocate) or oversubscribe (allocate more memory than is physically available) the memory in a way that scales gracefully rather than simply crashing the system. In fact, there is not even a requirement that the virtual and physical address spaces be the same size. In all of these ways, aside from some exceptions that we will discuss as they arise, each program can be blissfully oblivious to the physical implementation details or to any other programs that might be sharing the system.

The VM subsystem is also responsible for a number of other important memory management tasks. First of all, memory is allocated and deallocated regularly, and the VM subsystem must handle the available resources in such a way that allocation requests can be successfully satisfied. Naive implementations will lead to problems such as fragmentation in which an inefficient arrangement of memory regions in an address space leads to inaccessible and/or wasted resources. The VM subsystem must also gracefully handle situations in which memory is oversubscribed. It generally does so by swapping certain memory regions from memory to a backing store such as a disk drive. The added latency of going to disk generally results in a tremendous hit to performance, but some of that cost can be mitigated by a smart VM subsystem implementation.

Lastly, in a number of more recently emerging scenarios, memory will sometimes need to be migrated from one physical address to another. This can be done to move memory from one socket to another, from type of physical memory into another (e.g., DRAM to non-volatile memory), from one device to another (e.g., CPU to GPU), or even just to defragment memory regions within one physical memory block. VM provides a natural means to achieve this type of memory management.

1.2 ISSUES WITH MODERN VIRTUAL MEMORY

On architectures making use of VM, the performance of the VM subsystem is critical to the performance of the system overall. Memory accesses traditionally make up somewhere around one third of the instructions in a typical program. Unless a system uses virtually indexed, virtually tagged caches, every load and store passes through the VM subsystem. As such, for VM to be practical, address translation must be implemented in such a way that it does not consume excessive hardware and software resources or consume excessive energy. Early studies declared that address translation should cost no more than 3% of runtime [35]. Today, VM overheads range from 5–20% [10–12, 18–20, 22, 44, 66, 79, 88, 90], or even 20–50% in virtualized environments [17, 19, 32, 44, 67, 91].

However, the benefits of VM are under threat today. Performance concerns such as those described above are what keep VM highly relevant as a contemporary research area. Program working sets are becoming larger and larger, and the hardware structures and operating system algorithms that have been used to efficiently implement VM in the past are struggling to keep up. This has led to a resurgence of interest in features like superpages and TLB prefetching as mechanisms to help the virtual memory subsystem keep up with the workloads.

Furthermore, computing systems are becoming increasingly heterogeneous architecturally, with accelerators such as graphics processing units (GPUs) and digital signal processors (DSPs) becoming more tightly integrated and sharing virtual address spaces with traditional CPU user code. The above trends are driving lots of interesting new research and development in getting the VM abstraction to scale up efficiently to a much larger and broader environment than it had ever been used for in past decades. This has led to lots of fascinating new research into techniques for migrating pages between memories and between devices, for scalable TLB synchronization algorithms, and for IOMMUs which can allow devices to share a virtual address space with the CPU at all!

Modern VM research questions can largely be classified into areas that have traditionally been explored, and those that are becoming important because of emerging hardware trends. Among the traditional topics, questions on TLB structures, sizes, organizations, allocation, and replacement policies are all increasingly important as the workloads we run use ever-increasing amounts of data. Naturally, the bigger the data sizes, the more pressure there is on hardware cache structures like TLBs, and MMU caches, triggering these questions. We explore these topics in Chapters 3–5.

Beyond the questions of functionality and performance, correctness remains a major concern. VM is a complicated interface requiring correct hardware and software cooperation. Despite decades of active research, real-world VM implementations routinely suffer from bugs in both the hardware and OS layers [5, 80, 94]. The advent of hardware accelerators and new memory technologies promises new hardware and software VM management layers, which add to this already challenging verification burden. Therefore, questions on tools and methodologies that allow for disciplined formal reasoning of VM correctness becomes even more pressing going forward. We study these difficult correctness concerns in Chapter 6.

Finally, emerging computing trends pose even newer and more fundamental questions. For example, as systems embrace increasing amounts of memory, we must answer a fundamental question: VM was originally conceived when memory was scarce, so does it still make sense to use it? And if so, what parts of it are most useful looking ahead, and how must they be architected for good performance? Answering these questions in turn requires asking questions about the benefits of paging vs. segmentation, appropriate page sizes, page migration mechanisms and policies among sockets and among emerging die-stacked and non-volatile memory, and the right VM support for emerging hardware accelerators beyond even GPUs. We explore these highly timely topics in Chapters 7–9.

At the end of the book, in Chapter 10, we conclude with a brief perspective about where we see the field moving in the coming years, and we provide some thoughts on how researchers, engineers, and developers might find places where they can dive in and start to make a contribution.

CHAPTER 2

The Virtual Memory Abstraction

Before we dive into the implementation of the VM subsystem later in the book, we describe the VM abstraction that it provides to each process. This lays out the goal for the implementation details that will be described in the rest of this book. It also serves as a refresher for readers who might want to review the basics before diving into more advanced topics.

2.1 ANATOMY OF A TYPICAL VIRTUAL ADDRESS SPACE

We start by reminding readers of the important distinction between "memory" and "address space," even though the two are often used interchangeably in informal discussions. The former refers to a data storage medium, while the latter is a set of memory addresses. Not every memory address actually refers to memory; some portions of the address space may contain addresses that have not (yet) actually been allocated memory, while others might be explicitly reserved for mechanisms such as memory-mapped input/output (MMIO), which provides access to external devices and other non-memory resources through a memory interface. Where appropriate, we will be careful to make this distinction as well.

A wide variety of memory regions are mapped in the address space of any general process. Besides the heap and stack, the memory space of a typical process also includes the program code, the contents of any dynamically linked libraries, the operating system data structures, and various other assorted memory regions. The VM subsystem is responsible for supporting the various needs of all of these regions, not just for the heap and the stack. Furthermore, many of these other regions have special characteristics (such as restricted access permissions) that impose specific requirements onto the VM subsystem.

The details of how a program's higher-level notion of "memory" is mapped onto the hardware's notion of VM is specific to each operating system and application binary interface (ABI). A common approach is shown (slightly abstracted) in Figure 2.1. At the bottom of the address space are the program's code and data sections. At the top is the region of memory reserved for the operating system; we discuss this region in more detail below. In the middle lie the dynamically changing regions of memory, such as the stack, the heap, and the loaded-in shared libraries.

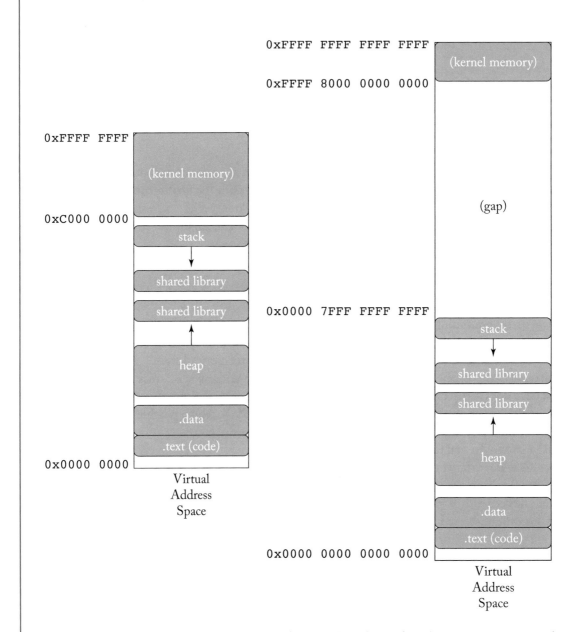

Figure 2.1: A cartoon of the memory space of a typical 32-bit and 64-bit process, respectively. (Figures not drawn to scale). See also Figure 2.3.

Traditionally, the stack was laid out at one end of the memory space, and the heap was laid out in the other, with both growing in opposite directions toward a common point in the middle. This was done to maximize the flexibility with which capacity-constrained systems (including even 32-bit systems) could manage memory. Programs that needed a bigger stack than heap could use that space to grow the stack, and programs that needed more heap space could use it for the heap instead. The actual direction of growth is mostly irrelevant, but in practice, downward-growing stacks are much more common. In any case, today's 64-bit applications typically have more virtual address space than they can fill up, so collisions between the stack and the heap are no longer a major concern.

While Figure 2.1 was merely a cartoon, Figure 2.2 shows a more concrete memory map of a prototypical 32-bit Linux process called /opt/test, as gathered by printing the contents of the virtual file /proc/<pid>/maps (where pid represents the process ID). Each line represents a particular range of addresses currently mapped in the virtual address space of the specified process. Some lines list the name of a particular file backing the corresponding region, while others—such as those associated with the stack or the heap—are *anonymous*: they have no file backing them. Of course, the memory map must be able to adapt dynamically to the inclusion of shared libraries, multiple stacks for multiple threads, and any general random memory allocation that the application performs. We discuss memory allocation in detail in Section 5.3.

Finally, a portion of the virtual address space is typically reserved for the kernel. Although the kernel is a separate process and conceptually should have its own address space, in practice it would be expensive to perform a full context switch into a different address space every time a program performed a system call. Instead, the kernel's virtual address space is mapped into a portion of the virtual address space of each process. Although it may seem to be, this is not considered a violation of VM isolation requirements, because the kernel portion of the address space is only accessible by a process with escalated permissions. Note that with the single exception of vDSO (discussed below), kernel memory is not even presented to the user as part of the process' virtual address space in /proc/<pid>/maps. This pragmatic tradeoff of mapping the kernel space across into all processes' virtual address spaces allows a system call to be performed with only a privilege level change, not a more expensive context switch.

The partitioning between user and kernel memory regions is left up to the operating system. In 32-bit systems, the split between user and kernel memory was a more critical parameter, as either the user application or the kernel (or both!) could be impacted by the maximum memory size limits being imposed. The balance was not universal; 32-bit Linux typically provided the lower 3 GB of memory to user space and left the upper 1 GB for the kernel, while 32-bit Windows used a 2 GB/2 GB split.

On 64-bit systems, except in some extreme cases, virtual address size is no longer critical, and generally the address space is simply split in half again. In fact, the virtual address space is so large today that much of it is often left unused. The x86-64 architecture, for example, currently requires bits 48-63 of any virtual address to be the same, as shown in Figure 2.3. Addresses

```
address             perms offset  dev    inode       pathname

08048000-08049000 r-xp 00000000 03:00 8312          /opt/test
08049000-0804a000 rw-p 00001000 03:00 8312          /opt/test
0804a000-0806b000 rw-p 00000000 00:00 0             [heap]
a7cb1000-a7cb2000 ---p 00000000 00:00 0
a7cb2000-a7eb2000 rw-p 00000000 00:00 0
a7eb2000-a7eb3000 ---p 00000000 00:00 0
a7eb3000-a7ed5000 rw-p 00000000 00:00 0
a7ed5000-a8008000 r-xp 00000000 03:00 4222          /lib/libc.so.6
a8008000-a800a000 r--p 00133000 03:00 4222          /lib/libc.so.6
a800a000-a800b000 rw-p 00135000 03:00 4222          /lib/libc.so.6
a800b000-a800e000 rw-p 00000000 00:00 0
a800e000-a8022000 r-xp 00000000 03:00 14462         /lib/libpthread.so.0
a8022000-a8023000 r--p 00013000 03:00 14462         /lib/libpthread.so.0
a8023000-a8024000 rw-p 00014000 03:00 14462         /lib/libpthread.so.0
a8024000-a8027000 rw-p 00000000 00:00 0
a8027000-a8043000 r-xp 00000000 03:00 8317          /lib/ld-linux.so.2
a8043000-a8044000 r--p 0001b000 03:00 8317          /lib/ld-linux.so.2
a8044000-a8045000 rw-p 0001c000 03:00 8317          /lib/ld-linux.so.2
aff35000-aff4a000 rw-p 00000000 00:00 0             [stack]
ffffe000-fffff000 r-xp 00000000 00:00 0             [vdso]
```

Figure 2.2: Contents of /proc/<pid>/maps for a process running on Linux (taken from Documentation/filesystems/proc.txt in the Linux source code).

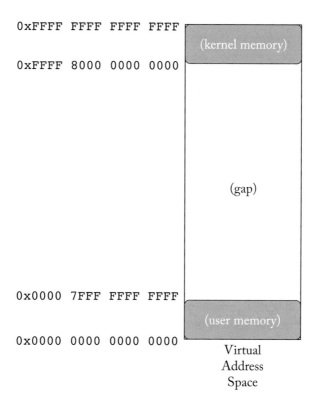

Figure 2.3: The x86-64 and ARM AArch64 architectures currently require addresses to be in "canonical form": bits 48–63 should always be the same. This leaves a large unused gap in the middle of the 64-bit address space.

meeting this requirement are said to be in "canonical form". Any accesses to virtual addresses not in canonical form result in illegal memory access exceptions. Canonical form still leaves 2^{47} bytes (256 TB) accessible, which is sufficient for today's systems, and the canonical form can be easily adapted to use anything up to the full 64 bits in the future if necessary. For example, the x86-64 architecture is already moving toward 57-bit virtual address spaces in the near future [56]. Choosing these sizes in a way is a delicate tradeoff between practical implementability concerns and practical workload needs, and it will remain a very important point of discussion in the field of VM for the foreseeable future.

One final line in Figure 2.1 merits further explanation: what is vDSO? The permissions (discussed below) also indicate that it is directly executable by user code. Note however that vDSO lives in the kernel region of memory (0xC0000000–0xFFFFFFFF), the rest of which is simply not accessible to user space. What is going on? vDSO, the virtual dynamically linked shared object, is a special-purpose performance optimization that speeds up some interactions between user and kernel code. Most notably, the kernel manages various timekeeping data structures that are sometimes useful for user code. However, gaining access to those structures traditionally required a system call (and its associated overhead), just like any other access into the kernel from user space. Because user-level read access to those kernel data structures posed no special security risks, vDSO (and its predecessor vsyscall) were invented as small, carefully controlled, user-accessible regions of kernel space. User code then interacts with vDSO just as it would with a shared library. Just as with the rest of kernel memory, this pragmatic workaround provides a great example of the various sophisticated mechanisms that go into defining, enforcing, and optimizing around memory protection in modern virtual memory systems.

We make one final point about the address space: the discussion above does not change fundamentally for multithreaded processes. All of the threads in a single process share the same address space. A multithreaded process may have more than one stack region allocated, as there is generally one stack per thread. However, all of the other virtual address space regions discussed above continue to exist and are simply shared among all of the threads. We discuss multithreading in more detail in Section 2.3.

2.2 MEMORY PERMISSIONS

One major task of the VM subsystem is in managing and enforcing limited access permissions on the various regions of memory. There are three basic memory permissions: read, write, and execute. In theory, memory regions could be assigned any combination of the three. In practice, for security reasons, pages generally cannot have read, write, and execute permission simultaneously. Instead, most memory regions are assigned some restricted set of permissions, according to the purpose served by that region.

Adding permissions controls explicitly into the VM system makes it easier for the system to reliably catch and prevent malicious behavior. Table 2.1 summarizes many common use cases. Memory regions used to store general-purpose user data are readable and sometimes writable, but are not executable as code. Likewise, memory regions containing code are generally readable and executable, but executable pages are generally not writable in order to make it more difficult for malware to take control of a system. This is known as the W^X ("write XOR execute") principle.

Table 2.1: Use cases for various types of memory region access permissions

Read	Write	Execute	Use Cases
Y	Y	Y	Code or data; was common, but now generally deprecated/discouraged due to security risks
Y	Y	—	Read-write data; very common
Y	—	Y	Executable code; very common
Y	—	—	Read-only data; very common
—	Y	Y	N/A
—	Y	—	Interaction with external devices
—	—	Y	To protect code from inspection; uncommon
—	—	—	Guard pages: security feature used to trap buffer overflows or other illegal accesses

Other permission types are less common, but do exist. Write-only memory may seem like a joke, but it turns to be the most sensible way to represent certain types of memory-mapped input/output (MMIO) interaction with external devices (see Section 7.3.3). Likewise, execute-only regions may seem strange, but they are occasionally used to allow users to execute sensitive blocks of code without being able to inspect the code directly. In the extreme, guard pages by design do not have any access permission at all! Guard pages are often allocated just to either side of a newly allocated virtual memory region in order to detect, e.g., an accidental or malicious buffer overflow. Due to the restricted permissions, any access to a guard page will result in a segmentation fault (Section 3.4) rather than a data corruption or exploit.

Many region permissions are derived from the segment in the program's object file. For example, consider the binary of a C program. The .data and .bss (zero-initialized data) segments will be marked as read/write. The .text (code) segment contains code and will be marked as read/execute. The .rodata (read-only data) segment will, not surprisingly, be marked as read only.

Some specialized segments of a C binary, such as the procedure linkage table (.plt) or global offset table (.got) may be marked read-or-execute. However, the dynamic loader in the operation system is responsible for lazily resolving certain function calls, and it does so by patching the .plt or .got sections on the fly. For this reason, and because of W^X restrictions, the dynamic loader may occasionally need to exchange execute permission for write permission in order to update the contents of the `.plt` or `.got` segments. This use of self-modifying code is subtle but nevertheless important to get right. We discuss the challenges of self-modifying code in Section 6.3.

Shared libraries generally have a structure which is similar (or even identical) to executable binaries. Many shared libraries are in fact executables themselves. When a shared memory is loaded, it is mapped into the address space of the application, following all of the same permission rules as would otherwise a apply to the binary. For example, a C shared library will follow all of the rules listed above for C executables.

Users can also allocate their own memory regions using system calls such as `mmap` (Unix-like) or `VirtualAlloc` (Windows). The permissions for these pages can be assigned by the user, but may still be subject to hardware-enforced restrictions such as W^X. The user can also change permissions for these regions dynamically, using system calls such as `mprotect` (Unix-like) or `VirtualProtect` (Windows). The operating system is responsible for ensuring that users do not overstep their bounds by trying to grant certain permissions to a region of memory for which the specified permissions are illegal.

2.3   MULTITHREADED PROGRAMS

The virtual address space of a process can also adapt to multithreaded code. For clarity, because terminology can differ from author to author, we again start with some definitions. A *process* is one isolated instance of a program. Each process has its own private state, including, most importantly for this book, its own isolated virtual address space. A *thread* is a unit of execution running code within a process; each process can have one or more threads. For the purposes of this book, we focus on threads which are managed by the operating system as independently-schedulable units. *User threads*, which are user code libraries which provide a thread-like abstraction, are not seen as separate threads by the VM subsystem, and so we do not consider them further. Likewise, we do not distinguish between fibers (cooperative threads) and preemptive threads; we leave this and other similar discussions for operating system textbooks.

Multithreading does not in itself change much about the state of a process' virtual address space abstraction. All threads in a process share the same virtual address space, along with most of the rest of the process state. Likewise, because they share the same virtual address space, the threads also all share a single page table. However, since each thread runs in a separate execution context, each thread does receive its own independent stack, as shown in Figure 2.4.

The stacks for all of the threads in a process are mapped into the same address space, and so every stack is directly accessible by every thread, assuming it has the relevant pointer. Sharing data on the stack between threads is generally discouraged as a matter of code cleanliness, but it is not illegal from the point of view of the VM subsystem. Just as with the stack of a single-threaded program, the stacks of a multithreaded program are usually limited in size so that they do not clobber other memory regions (or each other). Furthermore, guard pages may be inserted on either side of each stack to catch (intentional or unintentional) stack overflows.

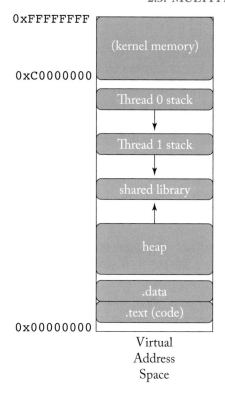

Figure 2.4: All threads in a process share the same address space, but each thread get its own private stack.

Distinct processes do not share any part of their address spaces unless they use one or more of the shared memory mechanisms described in the next section. However, an interesting situation arises during a process fork: when a parent process is duplicated to form an identical child process. As a result of the fork, the child process ends up with an exact duplicate of the virtual address space of the parent; the only change is that the process ID of the child will differ from that of the parent. At that point, the two virtual address spaces are entirely isolated, just as with any other pair of distinct processes. However, the physical data to which the two address spaces point is identical at the time of the fork. Therefore, there is no need to duplicate of the entire physical address range being accessed by the original process. Instead, most modern operating systems take advantage of this and make liberal use of copy-on-write (Section 5.1.2) to implement forks. Over time, as the two processes execute, their memory states will naturally begin to diverge, and the page table entries for each affected page will slowly be updated accordingly.

2.4 SHARED MEMORY, SYNONYMS, AND HOMONYMS

Virtual memory does not always enforce a one-to-one mapping between virtual and physical memory. A single virtual address reused by more than one process can point to multiple physical addresses; this is called a *homonym*. Conversely, if multiple virtual addresses point to the same physical address, it is known as a *synonym*. *Shared memory* takes this even further, as it allows multiple processes to set up different virtual addresses which point to the same physical address. The challenge of synonyms, homonyms, and shared memory lies in the way they affect the VM subsystem's ability to track the state of each individual virtual or physical page.

Shared memory is generally used as a means for multiple processes to communicate with each other directly through normal loads and stores, and without the overhead of setting up some other external communication medium. Just as with other types of page, shared memory can take the form of anonymous regions, in which the kernel provides some way (such as a unique string, or through a process fork) for two processes to acquire a pointer to the region. Shared memory can also be backed by a file in the filesystem, for example, if two different processes open the same file at the same time. It is also possible to have multiple virtual address ranges in the same process point to the same physical address; there is no fundamental assumption that shared memory mechanisms only apply to memory shared between more than one process.

From a conceptual point of view, synonyms, homonyms, and shared memory are straightforward to define and to understand. However, from an implementation point of view, the breaking of the one-to-one mapping assumption makes it more difficult for the VM subsystem to track the state of memory. Features which are performance-critical, such as forwarding of store buffer entries to subsequent loads, are generally handled in hardware. Some aspects of homonym and synonym management that might otherwise be handled by hardware are left instead to software to handle, meaning that the operating system must step in to fill the gaps left by hardware.

2.4.1 HOMONYMS

As described earlier, a homonym refers to a situation in which one virtual address points to more than one physical address. Figure 2.5 shows the standard example. Each virtual address within a single process maps to at most one physical address (zero if the virtual address is unmapped), and so homonyms arise when different processes reuse the same virtual address value. The virtual address spaces themselves are still distinct; it is only the numerical value of the virtual address that is being reused. The key challenge that homonyms present is that unless a process ID or address space ID of some kind is attached to the memory request, it can be impossible to tell which mapping should be used.

There are two basic solutions to the homonym problem. The first is to simply flush or invalidate the relevant hardware structures before any situation in which the two might otherwise be compared. For example, if a core performs a context switch, a TLB without process ID bits will have to be flushed to ensure that no homonyms from the old context are accidentally treated

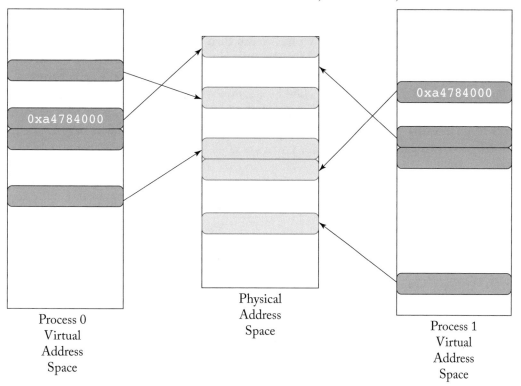

Figure 2.5: The virtual address 0xa4784000 is a homonym: it is mapped by different processes to different physical addresses.

as being part of the new context. The second is to attach a process or address space ID to each virtual address whenever two addressed from different process might be compared, and then to track the process or address space ID in the translation caching structures as well. For example, as we discuss in Section 4.5.3, some TLBs (but not all) associate a process ID of some kind with each page table entry they cache, for exactly this reason.

Importantly, the TLB is not the only structure which must account for homonyms. Virtually tagged (VIVT) caches, although not as common as physically tagged caches, would also be prone to the same issue and the same set of solutions. In fact, the solution of using process IDs can be incomplete for structures such as caches in which the data is not read-only. Even the store buffer or load buffer of a core might be affected: a store buffer that forwards data based on virtual addresses alone might also return confuse homonyms if it does not use either of the two solutions above. Any virtually addressed structure in the microarchitecture will have to adapt similarly.

2.4.2 SYNONYMS

Recall that a synonym is a situation in which more than one virtual address points to a single physical address. Figure 2.6 shows an example. Synonyms are the mechanism by which shared memory is realized, but as described earlier, synonyms are also used to implement features such as copy-on-write. The key new issue raised by synonyms is that the state of any given physical page can no longer be considered solely through the lens of any individual page table entry that points to it. In other words, with synonyms, the state of any given physical page frame may be distributed across multiple page table entries, and all must be considered before taking any action.

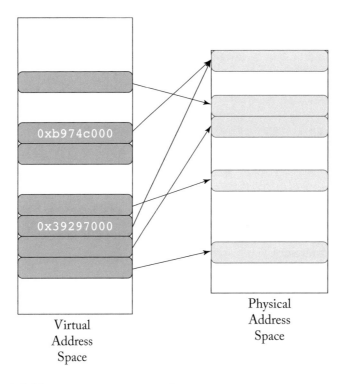

Figure 2.6: Virtual Addresses 0xb974c000 and 0x39297000 are synonyms: they both map to the same physical address.

One simple example of the problems posed by synonyms can be understood by considering the page table entry status bits. Suppose there is some synonym for a physical page which is currently marked clean (Section 4.5.2). Now, suppose one thread writes to that page. That will cause the page table entry for that virtual address to be marked dirty; however, it will not in general automatically update the page table entries for any synonym pages. Therefore, if a second thread were to query the status bits for its own page table entry for the same physical page frame, then it would think the page is clean. If the operating system is not careful to check all synonym page table entries as well, then it might erroneously think the page remains clean, and might overwrite it without first flushing any dirty data to the backing store.

Likewise, consider the process of swapping out a physical page frame to backing store. The first step in this process is to remove any page table entries pointing to that physical page frame, so that no process will be able to write to it while it is swapped out. In the case of synonyms, by definition, the reverse mapping (Section 5.3.5) now must be a multi-valued function. This means that even if a kernel thread already has a pointer to one page table entry for the given physical page frame, it must still nevertheless perform the reverse map lookup to find *all* of the relevant page table entries. This adds significant complexity to the implementation.

The microarchitecture is also directly affected by the synonym problem. First of all, as we will see in Section 4.2.2, cache indexing schemes can have subtle interactions with synonyms. Virtually tagged caches struggle to deal with synonyms at all, but low associativity VIPT caches also suffer from the fact that synonyms can map into different cache sets, breaking the coherence protocol. But caches are not the only parts of the architecture that are affected. Any structure that deals with memory accesses must also take care to ensure that synonyms are detected and handled properly.

Returning to the store buffer example from above: suppose a store buffer tags its entries with the virtual address and process ID of the thread that issued each store. If a load later came along to access a synonym virtual address, then by comparing based on virtual address and process ID alone, the load would miss on the store buffer entry and would instead fetch an even older value from memory, thereby violating the coherence requirement that each read return the value of the most recent write to the same physical address (see Chapter 6). A simple solution would be to perform all store buffer forwarding based on physical addresses; however, this would put the TLB on the critical path of the store buffer, which would make the TLB even more of a performance bottleneck than it would be if it were just attached to the L1 cache. A more common solution to this problem in high-performance processors is to have the store buffer speculatively assume that no synonym checking needs to be done, and then to have a fallback based on translated physical addresses later confirm or invalidate the speculation.

2.5   THREAD-LOCAL STORAGE

Some threading implementations also provide a mechanism for some form of thread-local storage (TLS). TLS provides some subset of the process' virtual address space which is (at least nominally) accessible only to that thread, and this in turn can sometimes make it easier to ensure that threads do not clobber each other's private data. TLS implementations can make use of hardware features, operating system features, and/or runtime library features; the division of labor tends to vary from system to system.

In an abstract sense, TLS works as follows. At the programming language level, the user is provided with some API for indicating that some data structure should be instantiated once per thread, within the local storage of that thread. At runtime, the TLS implementation assigns either a register (e.g., CP15 c13 on ARM, FS or GS on x86) or a base pointer to each thread. Any access to a thread-local variable is then transparently indexed relative to the base address stored in the register to access the specific variable instance for the relevant thread.

As a stylized example, consider the scenario of Figure 2.7. The threads share the same address space, but the pages marked "(TLS)" are reserved for the thread-local storage data. The user code of both threads will continue to use the virtual address 0x90ed4c000, but the thread-local storage implementation will (through software and/or through some hardware register) offset that address by that thread's base pointer (either 0x3000 and 0x9000 in the figure). This will result in the translation pointing to one of the two separate physical memory regions, as intended.

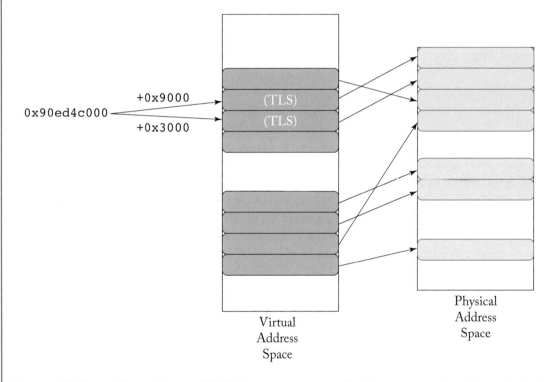

Figure 2.7: Thread-Local Storage (TLS) changes how virtual address are translated into physical addresses on a per-thread basis. In this example, both threads use virtual address 0x90ed4c000, but each thread adds its own distinct offset (0x3000 vs. 0x9000 here) prior to translation.

TLS highlights the role of the addressing mode of a memory access in a way that we have otherwise mostly glossed over as an ISA detail in this book. A virtual address may appear to the programmer to be one value, but it may through segmentation or register arithmetic become some modified value before it actually passed down into the VM memory subsystem. In the end, it is the final result of all of the appropriate segmentation calculations and register arithmetic operations that should be considered the actual VM address being accessed.

2.6   VIRTUAL MEMORY MANAGEMENT

Although programs are not directly responsible for managing physical memory, they do nevertheless perform many important memory management tasks within the virtual address space itself. Memory management tasks may come from the program explicitly (e.g., via `malloc`) or implicitly (e.g., stack-allocated function-local variables, or even basic allocation of storage for a program's instructions). In either case, management requests come from the programmer and/or the programming model, and they must ultimately trickle down through the operating system and the VM subsystem so that actual physical resources can be allocated to the program.

Programming languages generally provide a memory abstraction that sits even above the virtual address space. Thread-local variables that come and go as the code traverses each function in the program are often allocated on a first-in, first-out structure called the stack. Programmers may also dynamically allocate data meant to be persistent between function calls and possibly meant to be passed between threads. Such data structures are generally allocated in a random-access region commonly known as the heap. Programs may also have other regions of memory holding things like read-only constants compiled into the program. Of course, the details vary widely with each individual programming language.

In reality, the "heap" is just a generic name used to describe a region of memory within which a program can perform its dynamic random-access memory allocation. In fact, the notion of a heap often exists both at the language level and at the operating system level. In between the two generally sits a user-level memory allocation library. For example, C/C++ heap allocations using the `malloc` or `new` keywords are passed into the C library. The C library then either makes the allocation from within its pool(s) of already-allocated memory, or it requests more heap memory from the operating system through system calls such as `mmap` or `brk/sbrk`.

User-level memory management libraries are not technically a part of the VM subsystem, as they do not directly participate in the virtual-to-physical address translation process, nor do they manage physical resources in any way. Nevertheless, they play a very important role in keeping the VM subsystem running efficiently. For practical implementation reasons, system calls such as `mmap` or `brk/sbrk` generally only allocate VM at page granularity (or multiples thereof). They are also expensive, as calling them requires a system call into the operating system, and that in turn requires lots of bookkeeping to track the state of memory. User-level libraries generally filter out many OS system calls that might otherwise be needed by batching together small allocation requests or by reusing memory regions that have been freed by the program but not yet deallocated from the virtual address space. We explore user-level memory allocation libraries in more detail in Section 5.3.

The actual VM subsystem begins where the user-level memory management libraries stop. At the hardware's level of abstraction, the original purpose or source of the allocation request becomes mostly irrelevant; aside from differences in access permissions, all of the allocated memory regions become more or less functionally equivalent. With this in mind, the rest of this book focuses mostly on studying VM from the perspectives of the operating system and VM subsystem, decoupled from the particulars of any one program or programming language.

2.7   SUMMARY

In this chapter, we described the basics of the VM abstraction. We discussed how the VM abstraction presents each process with its own isolated view of memory, but we also discussed practical concessions such as canonical form and the mapping of kernel memory into the virtual address space of each process. In addition, we covered the permissions bits that govern access to each different memory region, we covered the tricker cases of synonyms, homonyms, and TLS, and then we jumped up one layer to discuss the types of programming models that add yet another layer of memory management on top of the VM subsystem itself.

The rest of this book is about the various sophisticated hardware and software mechanisms that work together to implement this virtual address space abstraction, as well as the research and development questions guiding the evolution of these mechanisms. There are many different aspects involved, both in terms of functional correctness and in terms of important performance optimizations which allow computers to run efficiently in practice. In the following chapters, we give an overview of the different components of the VM implementation, and then toward the end of the book, we explore some more advanced use cases in greater detail.

CHAPTER 3

Implementing Virtual Memory: An Overview

The benefits of VM cannot come for free. Address translation is generally on the critical path of every memory access, as in most cases the physical address must be determined before any actual data can be accessed. Since memory-intensive programs can spend as much as 30–40% of their instructions on memory accesses, address translation is critical to the performance of the entire program [10–12, 19, 20, 32, 43, 66, 86, 90]. To prevent address translation from becoming a bottleneck, many of the common-case operations used to perform address translation are implemented in dedicated hardware. The less performance-critical and/or more sophisticated management functions are left to the OS.

The exact hardware-software balance of the VM subsystem can vary from system to system. The key benefit of dedicated hardware is its improved performance and energy efficiency. The downside is the added area, design, and verification cost; chip vendors are generally hesitant to spend their transistor budget on hardware widgets unless those hardware widgets have convincingly proven their value over the corresponding software implementations. For some components of the VM subsystem, there is a near-universal consensus on their value. For example, nearly all processors today have translation lookaside buffers (TLBs) (Section 3.3). For other components, such as TLB shootdowns (Section 6.2), there is less consensus on the exact mechanism.

In this chapter, we present an overview of the operating system and hardware features that combine to implement the VM subsystem. We focus here on explaining the basic principles and mechanisms. The remaining chapters dive deeper into the details of each VM subsystem component and its design space.

3.1 A TYPICAL PAGING-BASED VIRTUAL MEMORY SUBSYSTEM

One of the important design decisions in a VM subsystem is the granularity at which address translation bookkeeping should be performed. Using relatively small granularity enables the most flexibility when making decisions about where to locate data and how to manage access permissions to each data structure in memory [32, 74, 86, 91]. However, fine-grained approaches also introduce the most overhead, as the amount of metadata that must be maintained to track

VM state grows enormously. Coarse-grained approaches require less overhead, but the inflexibility of only being able to manage data in large chunks can introduce its own challenges, such as memory fragmentation [82, 91].

Most modern VM subsystems use a strategy known as *paging*. Figure 3.1 shows a high-level overview. In this approach, the virtual address space is divided into *pages* and *page frames* (or simply *frames*). A page is a contiguous region of VM managed as a single unit, and a page frame is a contiguous region of physical memory managed as a single unit. Paging attacks the fragmentation problem by allowing pages to be swapped between primary and secondary storage.

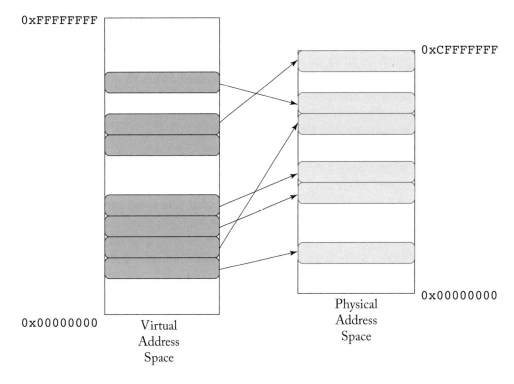

Figure 3.1: Virtual memory is a layer of abstraction that decouples user code from the details of the physical implementation of memory. This picture presents a cartoon view of virtual-to-physical mappings. Shaded boxes represent virtual pages or physical page frames. Arrows represent virtual-to-physical mappings for the given process. Note that mappings are not necessarily one-to-one.

Implementations today specifically use *demand paging*, which means that pages are brought into primary memory when (implicitly) requested by the user [42]. In other words, with demand paging, it is not the user's responsibility to migrate pages back and forth. Instead, the operating system automatically brings in the pages when they are accessed. Demand paging forms the foundation of nearly all VM systems today.

CPUs today use base page sizes of 4–64 KB, but this can vary. x86 and ARM systems today use 4 KB small pages. SPARC systems, on the other hand, use 8 KB base pages. In the past, the VAX 11 minicomputer of the 1970s used 512 B pages. Many CPUs today also make use of superpages to help relieve some stress on the VM subsystem. A superpage is any page size is any architecturally supported page size which is larger than the base page size. In contrast to base pages, which are on the order of kilobytes, superpages today may be as large as a gigabyte! Page size is a very important factor for architects to consider, and making the most efficient use of superpages remains an active area of research even today.

Page size is one key implementation detail that does bleed through the VM abstraction and affect basic VM functionality. Memory management through system calls like mprotect (Linux) or VirtualProtect (Windows) must be done at the granularity of the base page size. In almost every other way, user code can operate entirely unaware of the physical memory and of the VM subsystem, but restricting certain operations to take place at page granularity is yet another practical tradeoff that keeps the implementation costs manageable.

Figure 3.2 shows how the page size affects the address translation process. For a page size of 2^N bits, the lower N bits of the input virtual address represent an offset that lies within a single page, and so these low bits are passed through without modification. The remaining upper bits, representing granularities at least as large as the page size, are translated according to the mechanisms described in the rest of this chapter. The system might also check at this stage whether the input virtual address is in the proper canonical form (Section 2.1), if applicable on the system in question. Recall also that the number of input bit and output bits need not be the same, as the virtual and physical address spaces need not be the same size. The final physical address produced is the concatenation of the translated upper bits and the unmodified lower bits.

3.2   PAGE TABLE BASICS

In a paging-based implementation, the set of virtual-to-physical mappings for each process is stored in a data structure known as a *page table*. In its most basic form, a page table is simply a key-value data structure, with the key being the virtual address and the value being the corresponding physical address, with the caveat that page table lookups are performed at page granularity. There are many ways to instantiate the page table structure; we cover many of these below.

The basic unit of the page table is the *page table entry*, which stores all of the relevant information for one particular page's worth of virtual-to-physical address translation information. Figure 3.3 shows the page table entry format used on x86-64. Bits M through 12 (where M depends on the maximum physical address size for the implementation) store the physical page number for the virtual address in question; the remaining bits are either unused/reserved or store metadata about the page. All x86 pages have implicit read permission, but bits 63 and 1 indicate

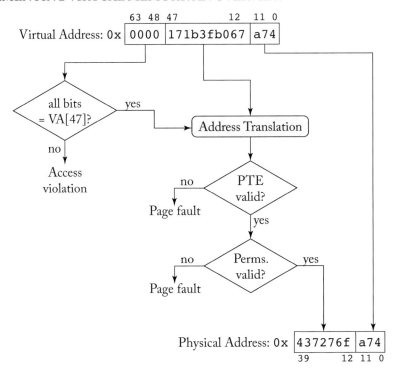

Figure 3.2: The process of translating a virtual address into a physical address, using an example with 64-bit virtual addresses, 48-bit canonical form, 40-bit physical addresses, and 4 KB pages (as with x86-64 and ARM AArch64).

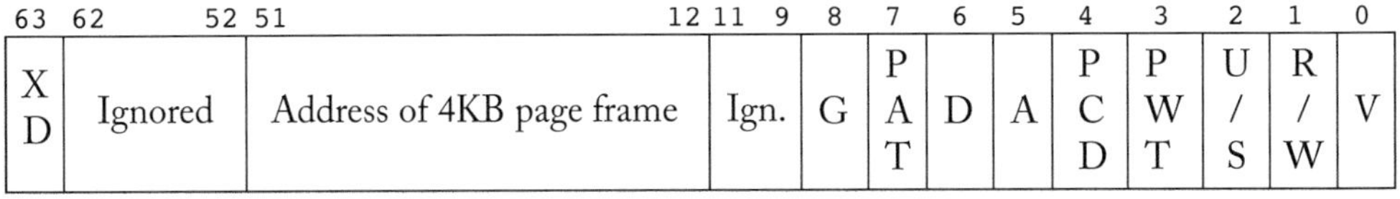

63 62		52 51		12 11 9	8	7	6	5	4	3	2	1	0
X D	Ignored	Address of 4KB page frame	Ign.	G	P A T	D	A	P C D	P W T	U / S	R / W	V

Figure 3.3: The page table entry format for base (4 KB) pages on x86-64 [55].

whether the page has execute and write permission, respectively. Bit 2 indicates whether the page is accessible only to supervisor mode (i.e., the OS or hypervisor). Bit 8 indicates whether the page is global: whether it can need not be invalidated upon performing a context switch (see Section 6.2). Bits 7, 4, and 3 store information about the cacheability of the memory region (Section 7.3.4). Finally, bits 6 and 5 indicate whether the page is dirty and accessed, respectively (Section 4.5.2).

The page table must be able to cover the entire virtual address space. Doing this naively would require an impractical amount of storage. Consider a typical system with 48-bit canonical form virtual addresses and 4 KB pages. On such a system, it would take 2^{48} B $/ 2^{12}$ B $\times 8$ B $= 2^{39}$ B $= 512$ GB of storage just for the page table alone! Using superpages could help trim this down, but only temporarily; for example, Intel's pending move to 57-bit canonical form addresses will only exacerbate the problem once again [56].

Fortunately, few applications use up the entire available virtual address space, and even fewer do so without making use of larger page sizes. Therefore, the most popular design today is variously known as the hierarchical, radix tree, or *multi-level page table*. Since it the most common design, we focus on the multi-level page table for the rest of this chapter, but we explain other page table designs in Section 4.1.

In a multi-level page table, translation proceeds in a number of steps. The process of traversing the levels of a page table is known as a *page table walk*, since it walks (i.e., chases pointers) through the different levels of the page table. A page table walk starts from the page table base address (labeled PT_BASE in Figure 3.4), which is unique to every process. The virtual address is divided into pieces, one for each level of the table. The topmost piece of the virtual address is used as an offset from the page table base address to produce a pointer to the first-level

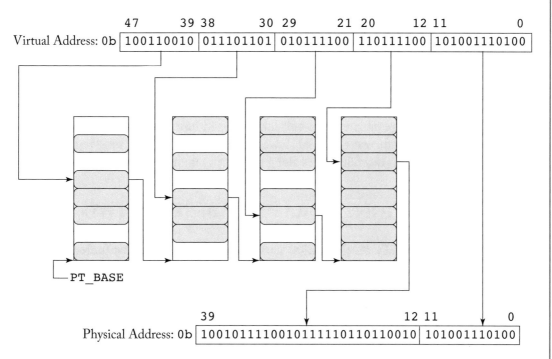

Figure 3.4: Multi-level page table.

page table entry. Entries in the first-level page table and in all other intermediate page tables are almost like normal page table entries, except that rather than pointing to physical page numbers, they instead point to the base address of the next level of the page table. This process repeats for every level in the table until reaching the end, and the final page table level's entry holds the physical page number.

At each step of the page table walk, each level-N page table entry also holds permission and status bits that provide metadata about the range of memory managed by that entry. For example, if a page may be mapped as read-only, a write access to that page is illegal and will result in a page fault (Section 3.4).

The number of levels in the page table varies across different architectures and according to the width of the address space. 32-bit x86 architectures use a two- or three-level page tables to translate 32 bits of virtual address into either 40 or 52 bits of physical address, respectively. 64-bit x86 and ARM architectures use four levels to translate 48 virtual address bits into 52 physical address bits, and Intel has already announced their plans to scale up to 57-bit virtual addresses using five-level page tables [56].

Multi-level page tables also play together very nicely with mixed page sizes [32, 74, 82]. An intermediate-level page table entry may also point directly to a physical address, rather than to a lower-level page table. Just as before, any remaining address bits are passed through unmodified. In this way, the translation is effectively performed at a page size which is larger than the original page size by a corresponding number of bits. This is shown in Figure 3.5.

Some page table entries may also simply not be present at all at the time of lookup. In fact, a major benefit of multi-level page tables is that they can be (and usually are) sparse. In other words, they do not necessarily have every single entry filled in. This sparsity is one of the keys to enabling page tables to scale to very large address spaces: large holes in the address space can simply be omitted from the page table. It is also true at each level: if no address that would be reachable by a level-N page table entry is currently mapped into the process in question, then neither that level-N page table entry nor any lower-level page table entries need to be allocated. In this way, the page table gracefully scales along with the memory usage of a process.

3.3   TRANSLATION LOOKASIDE BUFFERS (TLBS)

Since translation is generally on the critical path of every memory access (unless a cache is virtually tagged; see Section 4.2.2), frequently used address translations are cached in a small but low-latency hardware structure known as a *translation lookaside buffer* (TLB), as shown in Figure 3.6. The term "lookaside" refers to the fact that the TLB can (generally) be accessed in parallel with accessing the cache, rather than strictly before or strictly after. The benefit is clear: by storing commonly used translations in a fast structure near the processor, the need to perform most page table walks can be almost entirely eliminated. Instead, the processor directly searches the TLB for the physical address corresponding to the given virtual address.

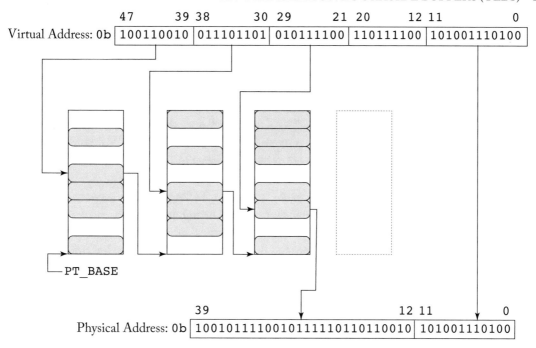

Figure 3.5: A multi-level page table with 4 KB base pages and 2 MB superpages.

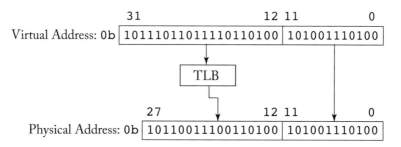

Figure 3.6: TLBs are small hardware caches of the page table.

TLBs provide a massive performance benefit to VM implementations; in fact, VM would likely be prohibitively expensive without them. Depending on the structure of the page table, memory bandwidth requirements would be instantly 2–5× higher. For example, x86-64 processors generally use a four-level page table, and so without TLBs, each memory access would become five (four levels of the page table, plus the access itself) [19]. Furthermore, the accesses are inherently sequential due to their pointer-chasing nature, making it difficult to hide the latency. Fortunately, TLBs have proven very successful at eliminating that overhead.

Of course, in order to remain fast, the TLB itself cannot be a very large structure, and so it can only hold a finite (and often somewhat small) number of translations [13, 88, 90]. Therefore, just as they do with data caches, many processors use a hierarchy of TLBs to cache translations. The lowest-latency (but consequently smallest) TLBs are placed directly next to each CPU, and larger (but higher-latency) TLBs are placed somewhat farther away. For example, for standard 4 KB pages, the Intel Skylake server-class architecture has an L1 TLB with 64 entries, giving it an addressable range of 4 KB × 64 = 256 KB of memory. The L2 TLB has 1,536 entries (for 4 KB pages), giving it an addressable range of 4 KB × 1,536 = 6 MB. For comparison, the smaller ARM Cortex A73 mobile architecture has 48 entries in its micro TLB (i.e., L1 TLB), and 1,024 entries in its main TLB (i.e., L2 TLB).

Like traditional on-chip hardware caches, modern TLBs are often set-associative: they are divided into ways and sets, just as many caches are. During any lookup, the set (the row) is determined from some subset of the input virtual or physical address. The lookup proceeds by searching all of the ways (or columns) for cache lines within that set, and it determines whether it has found a match by comparing the tag in the cache line against the tag derived from the rest of the input address. Set associativity allows caches to be made larger without excessively sacrificing latency.

TLB accesses are categorized according to their outcome, in much the same way as data cache accesses. If a TLB lookup finds a match in a TLB, it is known as a *TLB hit*. A TLB hit is the best case and the common case; it generally imposes no more than a few cycles of added latency. Otherwise, if the TLB lookup does not find a match, it is called a *TLB miss*. TLB misses can be expensive, as they require delaying the original access while the implementation searches the lower-level TLBs (if there are any), performs a page table walk, and/or traps into the operating system to handle the situation. An even costlier outcome is when the translation is not present in any TLB accessible to the core. In that case, either the hardware or the operating system must perform the page walk by traversing the page table in memory, and/or by taking a page fault. A great deal of effort has been put into making the most out of the limited storage available in the TLBs; we explore many of these issues throughout the rest of this book.

3.4   PAGE AND SEGMENTATION FAULTS

A *page fault* indicates either that the required translation information is not present in the page table at all or that the translation is present but with insufficient permission for the attempted access. A page fault occurs after a TLB miss: if the resulting page table walk indicates that the memory access cannot proceed based on the current state of the page table, it falls to the operating system to analyze the situation and to decide how to handle it. Some page faults are costlier than others, and page faults are occasionally used intentionally by the operating system to track accesses to some page of interest. To understand these situations, we start by describing various causes for page faults below.

Page faults are generally broken down into two categories: minor and major. A *minor page fault* means that the page frame is present in physical memory, but the translation was not present (or not set up with sufficient permissions) in the page table. This can happen, for example, if a page has been allocated but not yet accessed, in a scheme called *lazy allocation* (Section 5.1.1). A minor page fault can also happen if an already-mapped page frame is being given a second mapping. This happens frequently with shared libraries and inter-process shared memory. There also exist more advanced scenarios such as handling writes to copy-on-write pages; we will return to these later on.

A *major page fault* indicates that the desired data is not present in memory and hence that it must be fetched from backing storage, generally with a considerable latency overhead. If the requisite data has not yet been loaded from disk into memory, the major page fault will perform the transfer. Memory may also have been swapped out to disk due to limited memory capacity; accessing that data will incur a major page fault in order to bring it back into memory.

When a page fault does occur, the CPU usually assists the operating system in diagnosing the problem by providing some registers summarizing the situation. For example, a CPU might provide a some control/status registers (CSRs) indicating the faulting virtual address, whether the access was a read, a write, or an instruction fetch, whether it was user-level or kernel-level, and so on. The details of these bits are highly architecture-specific, and the procedures for how to handle page faults can be found in textbooks on operating system design. Generally, the OS will try to do what it can to find the requested physical data and map it into the virtual address space with the appropriate bits.

Lastly, a *segmentation fault* indicates that accessing the requested virtual address is simply illegal. The user may have requested access to a virtual address that was never allocated, or they may have tried to write (resp. execute code in) a page for which they have no write (resp. execute) permission. This commonly occurs due to programmer errors such as buffer overflows or trying to dereference a null or already-deallocated pointer. If this happens, the penalty is more than just a large latency overhead; the only safe thing to do in general after a segmentation fault is to simply terminate the process.

In any of the above cases, once a page fault has been successfully resolved, the memory access that originally triggered the page fault is *replayed*: it is re-executed under the assumption that the page table walk will now succeed. As the replay occurs, the CPU will again perform a page table walk to look up the newly inserted page table entry for the virtual address in question. This time, the page table walk will succeed, and so the translation can be inserted into the TLB and used to perform the memory access.

3.5   SEGMENTATION

The term "segmentation fault" is a bit of an anachronism that refers to segmentation: an alternative to paging-based VM. Segmentation is less popular nowadays than paging, but it was once the dominant scheme. On systems employing segmentation, the virtual address space is split into several logical segments. Typical segments are the code, data, stack, and heap. Each virtual address is formed by combining a segment number and an offset within the segment. The OS maintains a segment table to map the segment number to segment information, such as the base physical address of the segment, the limit of the segment, protection bits, and permissions bits (Section 2.2). Each memory access consults the segment table (or the segment table analog of a TLB) to determine the relevant physical address. A lookup failure results in a segmentation fault; the term persists even for paging-based VM.

The advantage of segments is the space efficiency that they provide. In particular, due to their large size, they enable small and fast TLBs, since they require relatively few translations. Compare segmentation, which requires one segment table entry to track an entire region, to paging, which requires a separate page table entry for each page. Clearly, the metadata overhead of segments can be much lower.

Nevertheless, the supposed space efficiency of segmentation over paging comes with two big problems. The first problem is that that segments are too large to enable fine-grained memory protection: a region of memory cannot be allocated or shared with another process with more restricted permissions than the segment it lives in. The second is that segments can be difficult to arrange in memory. The large size of segments can lead to memory fragmentation, where inefficient arrangement of the segments in memory can lead to wasted space. Making the problem worse is the fact that segments can change size dynamically. Segments cannot simply be packed together tightly, as doing so would prevent the first segment from growing up (or the second segment from growing down). However, spacing segments too far apart in memory would lead the memory in between the segments to be wasted. These issues make it difficult for an OS to manage segments in practice.

Due to these problems, modern systems use paging as the underlying mechanism to drive memory management. Systems with segments generally divide segments into constituent pages. For example, the x86 architecture uses a form of segmentation still today, but it does so partly for historical reasons, and the x86 use of segmentation is largely just an overlay on top of a paging-based VM subsystem anyway. Other systems without segmentation rely on paging entirely.

3.6   SUMMARY

In this chapter, we covered the basics of a typical VM subsystem implementation. We discussed multi-level page tables, which are the standard mechanism by which paging is implemented today. We also discussed TLBs, page faults, and segmentation faults. Most (but not all) VM systems today follow some variation of the basics presented in this chapter. Much of the research that continues to occur in the field of VM revolves around figuring out how to make the basic mechanisms described in this chapter run as efficiently as possible.

Having detailed the basic operation of VM, subsequent chapters will delve into its hardware and software implementation details. We will explore the design spaces that are of interest, and we will cover how some of these design points have largely converged, while others are changing quickly as architectures continue to adapt to the ever-changing needs of the computing world.

CHAPTER 4

Modern VM Hardware Stack

The VM subsystem is generally on the critical path of every instruction and data reference. Efficient support for VM is therefore important enough that most modern architectures are willing to dedicate hardware to make it as efficient as possible. In this chapter, we dive into some details of the design space of the hardware stack that makes up a modern VM subsystem. We cover both architectural details (such as the contents of the ISA-defined page table entry format) and microarchitectural details (such as the physical layouts of the TLBs).

4.1 INVERTED PAGE TABLES

Although the multi-level radix-tree page table is a common design, it is not the only possibility. Some architectures use different data structures for their page tables; each choice of data structure comes with different tradeoffs. This section explores some of these tradeoffs.

Some key factors that influence good page table design include the following.

1. Page table size: Modern systems employ a large virtual address space. Naive linear page tables are clearly inefficient uses of space; a 64-bit system with 48-bit virtual addresses and 4 KB pages would require 2^{36} page table entries, which at 8 B per PTE becomes 512 GB! Multi-level radix tree page tables are sparse and hence more efficient, but the upper levels of the page table are nevertheless overhead that alternative designs attempt to do away with.

2. Page table lookup: Even though TLBs are generally sized to reduce the frequency of TLB misses, page table lookups are unavoidable. Therefore, it is important to make page table searches as quick as possible. Multi-level radix-tree page tables require 3–5 memory accesses per page table walk, and as many as 35 in virtualized systems! (See Section 7.4.) Again, alternative designs often attempt to cut down on this pointer chasing and its high latency.

3. Efficient page table management: Finally, page tables must be efficient to maintain. That is, maintenance operations occur relatively often, and so keeping the VM subsystem efficient requires that adding, changing, and removing entries in the page table must be fast, as must the mechanisms for ensuring TLB coherence with page table hardware (Chapter 6).

One way to save page table space is to employ inverted page tables [57, 104, 115]. An inverted page table maintains one entry for every physical page in the system. The entry indicates which process uses this physical page, and which virtual page of that process maps to this physical page. As such, instead of having one multi-level radix page table per process, a single inverted page table is maintained for all processes in the system.

Figure 4.1 shows the basic operation of an inverted page table. The requested virtual address is split into the virtual page number (VPN) and page offset (Off). In ①–③, the PID and VPN are compared linearly to each inverted page table entry. In ③, the PID and VPN matches with the requested VPN. Therefore, the lookup reads the index value of the matching page table entry or PPN (0x18f1B), which is equal to the physical page number (PPN) in ④. The final physical address is formed in ⑤ by concatenating the PPN with the page offset.

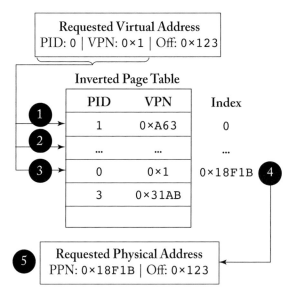

Figure 4.1: Inverted page table operation, involving a linear scan of the page table entries.

Finding the correct entry requires searching the inverted page table data structure. The naive option would be to employ a linear scan to look up the entire inverted page table. However, naive linear scans would clearly be expensive. Therefore, hash tables are usually built over the base inverted structure to speed lookups [115]. For example, hashed inverted page tables are used by the PowerPC architecture [54]. Typical hash functions employ an exclusive-or on the upper and lower bits of the virtual page number. Lookups first proceed by calculating the hash, and then by searching the table starting from the hash. Hashing cuts down dramatically on the number of page table entries that must be searched.

Since synonyms (Section 2.4.2) and/or different virtual page numbers may produce identical hash values, a collision resolution mechanism is used to let these mappings exist in the page table simultaneously. The basic solution is to start a collision chain: a list of alternative positions to search if the originally searched entry is a collision. A lookup first searches the entry at the hashed position. If it is a match, the lookup is complete. If it is a miss, the lookup then moves to the next entry in the collision chain and repeats. If the search reaches the end of the chain without finding a match, then the lookup results in a page fault.

Collision chains can be implemented in different ways. In classical inverted page tables, the collision chain resides within the table itself. Hence, when a collision occurs, the system chooses a different slot in the table into which it inserts the translation. This scheme is simple but not perfect: on TLB misses, if there are a lot of collisions, page table lookups can involve chasing a long list of pointers to find the desired translations.

Another option is that every time a translation is to be inserted in the page table, the inverted page table size is increased. In other words, to keep the average chain length short, the range of hash values is increased. In such a dynamically sized page table, it is necessary to explicitly include the page frame number within the page table entry. This increases the size of the inverted page table itself.

A more commonly employed approach is to use a Hash Anchor Table (HAT), pioneered in the PowerPC and UltraSPARC architectures [53, 58]. The HAT is an additional data structure that is indexed by the hash value and is accessed before the inverted page table. The hashed HAT entry points to the chain head in the inverted page table. Figure 4.2 illustrates the operation of a HAT-based inverted page table. On a TLB miss, the TLB faulting virtual page number is hashed in ①, indexing the HAT in ②. The corresponding HAT entry is loaded with a single memory reference; this entry points to the collision chain head in the inverted page table. This

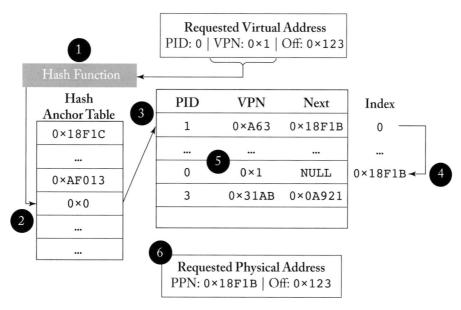

Figure 4.2: Hashed page table using a hash anchor table.

pipeline then generates a load memory reference using this pointer, from the inverted page table in ③. The virtual page number from that location is compared with the faulting virtual page. If the two match, the desired translation has been found and page table lookup can complete after this entry is filled into the TLB. If, however, there is a mismatch (see ④), the CPU loads the next translation in the chain. This process continues until the desired translation is found, or there is a page fault, as shown in ⑤. Ultimately, the physical address is calculated.

The key innovation with the HAT is the following: increasing the HAT size reduces average collision chain lengths without having to change the size of the inverted page table. Further, since the entries in the HAT are smaller than the entries in the inverted table, it is more efficient to increase the HAT to reduce the average collision chain length, rather than increasing the inverted page table size.

4.2   TLB ARRANGEMENT

Today's systems contain multiple TLBs and specialized caches scattered throughout a chip. TLBs located closer to the core generally aim to keep the common case of a hit as low-latency as possible, while TLBs located farther from the core generally aim to mitigate some of the cost of TLB misses, which can range in the tens to hundreds of clock cycles on a typical CPU [10, 11, 18, 19]. The goal of this section is to discuss the tradeoffs of various TLB arrangements. A VM system architect will have to evaluate the conditions below, in the context of the particular system design and workloads of interest, when deciding how to lay out the VM system for any new chip being built.

4.2.1   MULTI-LEVEL TLBS

As DRAM sizes and application memory footprints continue to increase over time, the demand for larger TLBs continues to grow accordingly. However, naively growing the size of the TLB would result in longer access latencies, and in most implementations this would in turn lengthen the critical path of every memory access. Therefore, for standard CPUs, where latency is critical, simply growing the TLB is not an attractive option.

Instead, rather than simply increasing TLB size, processors generally implement multiple levels of TLB in recent years [6, 9, 55]. Figure 4.3 shows how a multi-level TLB scheme might be arranged. If a translation lookup misses in the L1 TLB, then instead of triggering a page table walk immediately, the request instead searches the L2 TLB. An L2 TLB hit results in the translation being brought into the L1 TLB. An L2 TLB miss results in a page table walk. The same scheme can be extended for as many levels of TLB as are needed.

In order to keep the latency low, L1 TLBs are generally small, sometimes holding just a few entries. Last-level TLBs (those farthest from the core), on the other hand, might have as many as thousands of entries. Some TLBs are even implemented as set-associative structures in order to improve their efficiency even further. Today, many CPU vendors (e.g., Intel, AMD, and ARM) implement two levels of TLB per CPU. For example, for base pages, Intel's Kaby

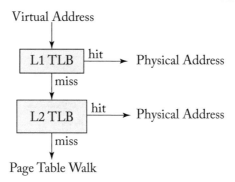

Figure 4.3: Multiple levels of TLB in a single CPU.

Lake processor implements 128-entry L1 instruction TLBs per core, 64-entry L1 data TLBs per core, and 1,536-entry unified L2 TLBs.

Another important TLB arrangement question is whether the last-level TLB should be private to each core or shared between multiple cores [20]. Per-core private TLBs suffer less contention and lower latency, as they would be physically closer to the core they serve, and this helps keep the cost of an L1 TLB miss lower. On the other hand, a shared last-level TLB would allow for a more efficient utilization of storage resources: any translations shared by multiple cores could be shared rather than duplicated, and this elimination of redundancy would allow the shared structure to provide a larger effective capacity. As shown in Figure 4.4, the viability of shared last-level TLBs therefore depends on whether the benefits of the increased last-level TLB hit rate outweigh the latency benefits of private last-level TLBs.

$$\text{Average TLB access latency} = \text{L1 hit rate} \times \text{L1 hit latency}$$
$$+ (1 - (\text{L1 hit rate})) \times \text{L2 hit rate} \times \text{L2 hit latency}$$
$$+ (1 - (\text{L1 hit rate})) \times (1 - (\text{L2 hit rate})) \times \text{L2 miss latency}$$

Figure 4.4: Calculating average TLB access latency. Shared vs. private last-level TLBs are a tradeoff between L2 TLB hit rate and L2 TLB hit latency.

Yet another important distinction is whether TLBs should be unified or split. Split TLBs store translation information for instruction and data memory separately, while unified TLBs store both together. Keeping instruction and data TLBs separate ensures that access to each by the relevant parts of the pipeline can be fast and uncontended. This is important because each may be on the critical path of instruction execution. Lower-level TLBs tend to be unified; since the latency of the lower-level TLB is not on the critical path of a hit, it makes more sense to use the space as efficiently as possible than it does to worry as much about contention.

4.2.2 TLB PLACEMENT RELATIVE TO CACHES

Conceptually, there are four separate placement strategies for modern TLBs relative to the caches, as summarized in Figure 4.5 and described below.

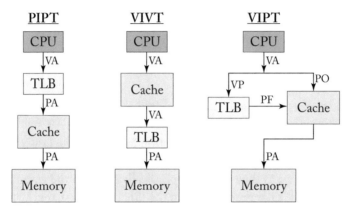

Figure 4.5: TLB placement strategies. On the left, we show physically indexed and physically tagged caches, where the TLB must be looked up before the cache and memory are looked up. In the center, we show virtually indexed and virtually tagged caches, where the TLB can be looked up before memory access, but caches be accessed using purely the virtual address. On the right, we show virtually indexed and physically tagged caches, where the TLB lookup and cache set index portions are overlapped.

Physically Indexed, Physically Tagged (PIPT) Caches: In this approach, virtual-to-physical translation is performed before caches are looked up. Figure 4.5 shows that the bits used to index PIPT caches are from the physical address, as are the bits used for tag match. Since code operates using virtual addresses, PIPT caches require TLB lookup to take place before cache lookup.

The primary benefit of PIPT is simple cache management. Since PIPT caches are physically addressed, they operate seamlessly in the presence of multiprogramming, and can easily snoop on cache coherence requests which also generally operate using physical addresses. Unfortunately, this benefit comes at the cost of performance. Since PIPT caches require TLB lookups to complete before cache lookups can even begin, TLBs must be fast. Consequently, TLBs cannot scale easily, increasing TLB misses, and hence degrading performance overall. As a result, vendors typically implement only L2 caches and LLCs as PIPTs [13]. L1 caches are not typically implemented, outside of simple embedded systems, as PIPT.

Virtually Indexed, Virtually Tagged (VIVT) Caches: VIVT caches obviate the need for TLB lookup completion prior to cache lookup [13, 28, 49, 68, 117]. Therefore, they permit TLBs to be scaled to larger sizes without compromising cache lookup times. Figure 4.5 shows that TLBs need to be looked up only after a virtually addressed cache. In fact, on a VIVT cache hit, the TLB may not need to be accessed at all

While VIVT does potentially improve performance, it introduces a host of issues that require complex and energy-intensive hardware structures to work around efficiently. For one thing, the fact that the TLB is not consulted on a cache hit means that a VIVT cache will not easily be able to detect and raise an exception on a memory access violation. In other words, an illegal access to an unallocated VM address or to an address that does not give permission for the access type in question will not be caught at the time of the access. Instead, such an illegal access would simply either never be trapped, or it would only be trapped once the line holding the illegally accessed virtual address is flushed from the cache. By then, the issuing instruction may have long since retired. One solution is to access the TLB even on a VIVT cache hit. This would have to be done before the access in question retires, but not necessarily before it executes; the latter ensures that it can be done efficiently.

Additionally, since caches are now tagged with virtual addresses, supporting multi-programmed workloads becomes challenging. Since different processes often use the same virtual addresses to refer to different physical addresses, and hence to different data, isolation may be violated. One approach to solving this problem is to add a process identifier (PID) or address space identifier (ASID) field to each cache line, as described in Section 4.5.3. Unfortunately, this approach increases the cache storage by requiring additional PID entries. In addition, TLBs maintain permission information that indicates whether a page is readable, writeable, executable, etc. Since TLBs are not looked up before VIVT cache lookup, permission information must be embedded in the cache line and looked up in parallel with tag and PID match.

Even within a single process, it is possible for different synonym virtual addresses to map to the same physical address. Synonyms are very difficult to manage in VIVT caches, since two distinct cache lines must now map the same data, and in fact the same physical address may even be mapped into two different sets! Decades of research by academia and industry have proposed several mechanisms to mitigate the problems of address synonyms, using a combination of ancillary structures that identify synonyms, cache allocation policies that identify synonyms and ensure that only one copy exists in the cache, and maintaining back-pointers and book-keeping structures to track the presence of synonyms [117]. We discuss synonyms in more detail in Section 2.4.2.

Due to all of the above complexities, VIVT caches are not extremely common in general-purpose processors in spite of their latency benefits.

Virtually Indexed, Physically Tagged (VIPT) Caches: VIPT caches present a compromise between high TLB hit rates and the benefits of physical addressing in caches [13, 105]. The key insight is that cache lookup is actually made up of two parts: set selection using index bits and tag match using tag bits. If the index bits are from the virtual address, the TLB lookup can be performed in parallel with perform cache set selection. Tag match proceeds only after set selection completes. As Figure 4.5 shows, with VIPT caches, the virtual page number is used to look up the TLB, while the page offset indexes the cache in parallel. Subsequently, cache tag match proceeds with a comparison to the physical address formed by concatenating the physical page extracted from the TLB with the untranslated page offset bits. Note that in order to perform cache set selection in parallel with TLB lookup, the cache index must be made up of bits that do not require translation. In the VIPT scheme, this means that index bits must come from the page offset.

The primary benefit of VIPT is that it completely hides the latency of the TLB access by overlapping it with the start of the cache access. At the same time, because VIPT caches use physical tags, the problems of multiprogramming and synonyms are circumvented. The drawback of VIPT, however, is that it imposes limits on the number of sets supported by the cache. This is because the cache index bits must be extracted from the page offset. For example, in a system with 4 KB pages, the page offset is made up of 12 bits. If the cache uses 64 byte cache lines, six of those 12 bits are used for the cache block offset, and hence only six bits remain for the index. This means that systems with 4 KB pages (e.g., ARM, x86) can support VIPT caches that have at most 64 sets. In turn, since the number of sets cannot grow, L1 VIPT cache associativity is generally directly proportional to the cache size, and the implementation costs of large highly associative caches generally mean that the L1 cache will stay relatively small. Systems with 4 KB base pages often maintain L1 caches on the order of 32 KB, as the resulting 8-way set associativity (32 KB total / 64 B lines / 64 sets = 8 ways) presents an appealing balance between the performance benefits and the implementation costs of different degrees of associativity.

If the above associativity guidelines are violated, then subtle problems with synonyms can arise once again. Consider what would happen if an architecture tried to implement a VIPT L1 cache with even less associativity than described above. Doing so would require more bits for the set index, and these bits would have to come from the virtual tag. This in turn would lead to one the same problems that VIVT caches face: that a single physical address might be mapped into multiple different sets by two different synonyms. In fact, this happened on certain ARMv6 cores, and it required the OS to implement a form of *page coloring* as a workaround [25]. At a high level, page coloring is a software scheme by which the memory allocator "colors" each page according to the position it would take in the cache. Tracking synonyms on low-associativity VIPT caches therefore requires OS support to ensure that all synonyms are in fact mapped with the same color. We describe page coloring in more detail in Section 5.3.4.

Physically Indexed, Virtually Tagged (PIVT) Caches: The last potential design option is to create a cache that is physically indexed and virtually tagged. However, if the physical address is known at the time of indexing, it is also known at the time of tag comparison, and so there is no reason to use virtual tags and all of their associated problems. Hence, PIVT caches are not used in practice.

4.3   TLB REPLACEMENT POLICIES

A TLB replacement policy is an algorithm that decides how and when to evict TLB entries from a full or almost-full TLB (or TLB way in a set associative TLB) in order to make room for new entries. Note that TLB replacement policies are distinct from cache replacement policies and the software replacement policies employed by the OS to determine which pages to evict to backing store; see Section 5.2 for details on the latter.

One possible TLB replacement policy is called least-recently used, or LRU. This is analogous to the LRU policy in caches: when a translation needs to be evicted, the least-recently accessed line is chosen as the victim. However, while the OS can employ sophisticated LRU (or approximate LRU) replacement, it is much harder to implement LRU in TLB hardware, even as an approximation, as the TLB has to abide by tight timing constraints. As a result, many modern L1 TLBs implement policies as simple as randomized replacement rather than LRU or its variants [20, 81, 108]. L2 TLBs generally have more relaxed timing constraints and hence can more easily afford to implement policies like FIFO replacement, pseudo-LRU, or LRU [20].

TLB replacement policies are not as widely studied in the literature as cache replacement policies, as the latter will generally have a much larger effect on performance. That being said, it is not yet clear how best to scale TLBs to some of the more advanced use cases we describe later in this book, and we expect that both implicit and explicit TLB management decisions such as these will have an important role to play in the VM systems of the future.

4.4   MULTIPLE PAGE SIZES

In order to balance memory management flexibility with TLB coverage, many modern OSes and architectures maintain support for multiple page sizes concurrently, as shown in Figure 4.6. For example, on x86 systems, superpages (2 MB and 1 GB pages) are used to increase TLB hit rates, as they cover a much greater portion of the address space with a single entry compared to the smaller base 4 KB pages. However, small 4 KB pages provide fine-grained page protection.

As we have discussed, TLBs are commonly set associative and often fully associative. Full associativity provides the best hit rates but at the highest implementation cost. Intermediate levels of associativity often strike a more appealing balance. However, set-associative TLBs paired with VIPT caches cannot (easily) support multiple page sizes. This is because, on lookup, TLBs need the lower-order bits of the virtual page number to select a TLB set. However, the location of those bits is not known unless the page size is also known, and the page size is not known until the TLB lookup has been successfully performed. This presents a chicken-and-egg problem, where the page size is needed for TLB lookup, but lookup is needed to determine page size. In general, industry and academia have responded in two ways.

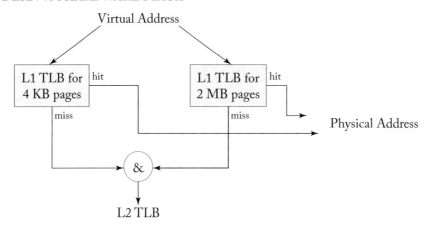

Figure 4.6: Multiple parallel TLBs for multiple page sizes. The TLBs are not necessarily the same size, and may be looked up serially or in parallel, depending on the particular design.

The first approach, used by most processor vendors today, uses split (or partitioned) TLBs, one for each page size [32, 86]. This side-steps the need for page size on lookup. A virtual address can look up all TLBs in parallel. Separate index bits are used for each TLB, based on the page size it supports, e.g., the set indices for split 16-set TLBs for 4 KB, 2 MB, and 1 GB pages are bits 15-12, 24-21, and 33-30, respectively. Two scenarios are possible. In the first, there is a hit in one of the split TLBs, implicitly indicating the translation's page size. In the second, all TLBs miss, in which case the access is treated as a normal TLB miss.

Unfortunately, while split TLBs are relatively simple, the parallel probes do waste energy, since a translation can only exist in one TLB. They also often underutilize TLBs, as the hardware resources cannot be flexibly allocated according to the page sizes actually in use at any given time. In other words, if the OS allocates mostly small pages, superpage TLBs remain wasted. On the other hand, when OSes allocate mostly superpages, performance may be (counter-intuitively) worsened because superpage TLBs (which are often smaller) thrash while small page TLBs lie unused [32, 43, 86]. Section 8.2 describes some more advanced alternatives.

4.5   PAGE TABLE ENTRY METADATA

The contents of page table entries themselves can also vary from architecture to architecture. All TLBs provide physical address information, but architectures can vary in the amount of metadata that is associated with those entries. In this section, we describe this metadata in more detail.

4.5.1 PERMISSION INFORMATION

As discussed in Section 2.2, page table entries and TLBs store information about access permissions to each region in the virtual address space. However, the specifics can vary from architecture to architecture. Some architectures provide a specific read permission bit, while others may simply assume that any memory region with a valid translation mapping is implicitly readable. Likewise, some architectures may provide W^X protection, while others may leave such decisions up to the operating system. Finally, some architectures may provide user and/or supervisor bits which gives permission to access a page only to a user process or to the operating system/hypervisor, respectively.

4.5.2 ACCESSED AND DIRTY BITS

In addition to virtual-to-physical page translations and permission information, page tables also maintain information that is vital for good OS page replacement decisions (Section 5.2.2). Two such pieces of information are encapsulated in each page table entry's access and dirty bits. We discuss each in turn.

1. Accessed bits are used by the system to mark pages that have been accessed in the recent past. The OS uses this information to identify cold pages that have not been accessed recently, as these prime candidates for eviction [2, 3]. Therefore, on every load and store operation, the accessed page's page table entry must have its accessed bit set. On modern CPUs, the page table walker is responsible for setting the accessed bit [19, 80, 94]. In other words, when there is a TLB miss and a page table walk is performed, the page table walker hardware identifies the desired page table entry and sets the accessed bit in it. Only once the accessed bit is set does the entry get filled in the TLB. Note that TLBs themselves do not generally maintain accessed bits within each entry because it is unnecessary—the accessed bits for all translations in the TLB must already have been set in the page table.

   Once the accessed bit is set, the only way it can be reset is by the OS. OSes typically use the accessed bit to approximate LRU by periodically resetting accessed bits to see whether the accessed bit is again set during the continued execution of the program [61, 75]. Those translations that see their accessed bits set again likely represent the hot pages in the application's memory footprint. Translations whose accessed bit are not set within a sufficiently long window are considered cold and are chosen by the algorithms described in Section 5.2 as leading candidates for replacement.

2. Dirty bits are used to identify memory-resident pages with data that needs to be written back to the stale backing-store [80, 94]. Much like dirty cache lines, this bit is set when a store instruction writes to a page. Unlike accessed bits, dirty bits are usually maintained in TLBs. Suppose that the TLB is looked up on a store instruction, and that the translation is found. Despite the fact that we seemingly have a TLB hit, if the dirty bit of the translation is set to zero, the page table walker must nevertheless be engaged to look up the TLB

entry's corresponding page table entry and set the dirty bit there. (Software cannot access the TLB contents directly, so dirty bits in the TLB simply filter out repeated setting of the bit in the page table, thereby saving bandwidth.) Once this is done, the dirty bit in the TLB is also set. Naturally, TLB misses proceed similarly, since the page table walker is engaged anyway.

While accessed bits are used mainly for performance reasons, dirty bits are in fact used to ensure proper functionality. If a dirty page is marked for eviction, then the data must be written back to the backing store before the physical memory allocated for that page can be evicted. Otherwise, the data in that page will be lost. Clean data, on the other hand, can be evicted without being written back, as the unset data bit indicates that nothing in the page had been modified anyway.

4.5.3   ADDRESS SPACE IDENTIFIERS AND GLOBAL BITS

Historically, TLBs did not track information about the context associated with each TLB entry, and this implied that OSes had to flush the TLB contents entirely on every context switch. This added up to a heavy cost: context switches occur thousands of times per second, leading to frequent TLB flushes. On a modern CPU, TLB flush operations generally take on the order of 5–20 microseconds (or more for bigger TLBs), plus subsequent memory operations now suffer TLB misses even in the presence of locality [33]. In general, wide-scale TLB flushing is known to cause as much as 10% performance degradations on ARM and x86 architectures [7, 89, 95, 110, 113].

Fortunately, many architectures have added hardware support to reduce the frequency of TLB flushes. TLBs maintain two fields: first, address space identifiers (ASIDs), and second, a global bit. Each process operates with its own ASID. When a process executes and fills TLB entries, the ASID is marked appropriately to ensure that the filled translation is associated only with the current process. TLB lookups check the ASID field and only when it matches the ASID of the executing process (generally set in some processor control register) can the entry experience a hit. Consequently, when there is a context switch, there is no need to flush the TLB. The new process will be unable to use TLB entries from the prior process' execution as there will be an ASID mismatch.

Note, however, that TLB ASIDs are not necessarily the same as OS process IDs. Specifically, the former tends to have fewer bits. For example, current x86 hardware provides 12 bits for TLB ASIDs, while Linux uses a 32 bit `pid_t` to track process IDs. Therefore, it falls to the operating system to track the mapping between hardware ASIDs and software context information. In some cases, this extra bookkeeping requirement is sufficiently burdensome that even today ASIDs are not always used even when they are available in the ISA.

In addition to ASIDs, the global bit aids performance too. The global bit is used to identify translations that are global to all processes [38]. This is useful generally for portions of the address space that are in use by the kernel. TLB entries with set global bits can be looked up by all processes; there need not be an ASID match to hit on a global TLB entry. Likewise, global TLB entries do not need to be flushed on context switches, even in systems without ASIDs, as they remain valid in the new context as well.

4.6   PAGE TABLE WALKERS

Despite innovations in TLB design that boost hit rates, misses can be unavoidable. Recall the basic capacity argument of Section 3.3: an Intel Skylake L2 TLB with 1,536 entries and 4 KB pages has an accessible range of only 6 MB, much smaller than the working set of a typical application today. Therefore, processor vendors have also designed progressively higher performance TLB miss handling mechanisms over several generations. We now present an overview of these mechanisms.

4.6.1   SOFTWARE-MANAGED TLBS

Page tables can be walked using either hardware or software support. In the early days of VM, software-managed TLBs using purely OS support for page table walks were the norm. In this approach, a TLB miss triggers an interrupt to the OS. The OS then runs an interrupt handler which performs the page table walk. Once the page table walk is completed, and the TLB is filled via a dedicated instruction, and control is passed back to the user-level process. This approach was popular among early SPARC, MIPS, and ARM processors through the 1990s and early 2000s [20, 22, 29, 57, 59, 81].

While conceptually simple to implement, software-managed TLBs suffer from poor performance. Every TLB miss requires a context switch to the OS. The pipeline must be flushed entirely. An interrupt handler must then be executed, polluting the contents of the on-chip hardware caches and large predictors (e.g., the branch predictor, memory disambiguation predictors). Then, another pipeline flush ends the interrupt handler, and finally, control is relinquished to the user process. Due to this heavy penalty, apart from some niche embedded CPUs, processor vendors have supplanted software-managed approaches with hardware-managed TLBs.

4.6.2   HARDWARE-MANAGED TLBS

The key benefit of a hardware-managed TLB is to mitigate the cost of requiring a trap into the OS to walk the page table. In other words, a TLB miss does not trigger an OS interrupt and the execution of an interrupt handler. Instead, CPUs maintain a hardware finite state machine called a hardware page table walker. After a TLB miss, page table walkers inject memory references for the page table walk directly into the pipeline. Aside from the fact that it is being performed by hardware, the walk itself proceeds as normal. Once the translation is identified, it is inserted into the TLB. The original memory reference is then replayed and execution continues.

There are several performance benefits from performing page table walks purely in hardware. First, the need for interrupts, pipeline flushes, and cache and branch predictor pollution are completely obviated. Second, since the pipeline continues executing instructions from the program, it is possible to overlap the page table walk with useful work, as long as the core's instruction scheduler can find independent instructions to issue [12, 20, 43]. This is preferable to software-managed TLBs, where the entire latency of the page table walk sits on the critical path of execution of every instruction subsequent to the miss.

The vast majority of hardware page table walkers access page tables using physical memory addresses directly, rather than VM addresses. This obviates the need for a chicken-and-egg problem of requiring virtual-to-physical address translation for the page tables, which provide virtual-to-physical address translations themselves. Software (i.e., the operating system), however, continues to access the page table using virtual addresses in exactly the same way as it would when accessing any other memory range.

Of course, the page table walker needs to know the base address of the current context's page table, and so architectures generally provide a control register which is configured by the OS as part of each context switch. For example, x86 systems use the CR3 register to store this information. In contrast, ARM uses two registers. The first, TTBR0, maintains a pointer to per-process page tables. The second, TTBR1, maintains a pointer to the page table entries that are globally shared among processes, and usually correspond to OS kernel structures. Like x86 systems, ARM systems maintain physical memory addresses in these registers.

The disadvantage of hardware page table walks is that the walker's hardware finite state machine is hard-wired to operate with a specific page table organization and lookup approach. Hardware page table walkers reduce the potential flexibility of changing page table organizations. Nevertheless, processor vendors deem the performance benefits of hardware page table walkers to more than make up for this loss in flexibility. Therefore, almost all chips over the last few generations use hardware page table walkers.

In fact, processor vendors are continuing to innovate on hardware page table walker design today; for example, Intel and AMD have begun integrating multiple (two or four) page table walkers per CPU today [111]. As the CPUs are designed to accommodate more in-flight instructions and wider instruction windows, is likely that several in-flight loads and stores might simultaneously suffer TLB misses. Multiple page table walkers can handle these TLB misses in parallel, improving performance.

4.6.3   MMU CACHES

TLBs are not the only structures used to cache address translation information. As modern workloads continue to grow well beyond the reach even of last-level TLBs, the cost of a TLB miss is becoming increasingly critical to performance. In an effort to speed up the latency of a TLB miss, some processor vendors (e.g., Intel, AMD, and ARM) design structures that cache page table entries from earlier levels of a multi-level radix page table. These small per-core hardware page table caching structures are known generically as Memory Management Unit (MMU) caches [10, 17, 18, 115]. TLBs cache PTEs from the last level of the page table level, while MMU caches, in comparison, store L4, L3, and L2 PTEs.

MMU caches are accessed on TLB misses during the hardware page table walk process. At each level of the walk, the page table walk state machine first checks whether the requested entry is present in that level's MMU cache. If there is a hit, then the walker can proceed to the next level of the walk immediately. If not, the walker accesses memory (possibly through the caches, depending on the implementation) in order to find the desired entry. In the best case, if the walk hits in the MMU cache at each level of the page table, almost all of the latency of the walk will have been eliminated.

The motivation for building MMU caches is that it makes sense to dedicate caching for upper-level PTEs because they map larger portions of the address space. For example, in a standard x86-64 four-level paging scheme, each successive level of the page table covers 512 GB, 1 GB, 2 MB, and 4 KB, respectively. Earlier levels especially are highly likely to be reused, and hence it makes sense to prioritize keeping those entries easily accessible to the page table walker, rather than having them fight for space with all of the other data stored in the caches.

There are many possible variants of the basic MMU cache idea. Intel uses Paging Structure Caches (PSCs), which are indexed by parts of the virtual address [10, 18]. Separate PSCs are maintained for each page table level; L4 entries are tagged with the L4 index, L3 entries are tagged with both L4 and L3 indices, while L2 entries are tagged with L4, L3, and L2 indices. On a TLB miss, all PSC entries are searched in parallel, and the longest match (if any) is used as the starting point for the rest of the page table walk. As a result, the walk can be completed with fewer actual memory references, saving latency.

Figure 4.7 shows how PSCs work. When looking up virtual address 0x5c8315cc2016, even if the processor sees a TLB miss, it would hit in each of the L4, L3, and L2 PSCs. The longest match (and hence the one that saves the most memory references) is the L2 PSC entry, and so the walk proceeds from that entry. Only the L1 entry at offset 0c2 from base address 0x508 needs to be accessed in memory.

Another option for building MMU caches is the AMD Page Walk Cache (PWC) [10, 17, 18]. Unlike PSCs, PWCs are simply dedicated PIPT caches for each page table level. Therefore, each levels must be looked up sequentially, but possibly at much lower latency (with a MMU cache hit). MMU caches remain an active area of research at present. Processor vendors are continuing to innovate on MMU cache design, lowering their access times to the range of 8–15 cycles today. As a result, they provide an important optimization that keeps the VM system working as efficiently as possible.

4.6.4   TRANSLATION STORAGE BUFFERS

We conclude this chapter by discussing software caching support for address translation that goes beyond traditional TLBs and MMU caches. SPARC's Translation Storage Buffer (TSB) is an example of such software support [20, 22, 57]. TSBs are data structures used to speed up

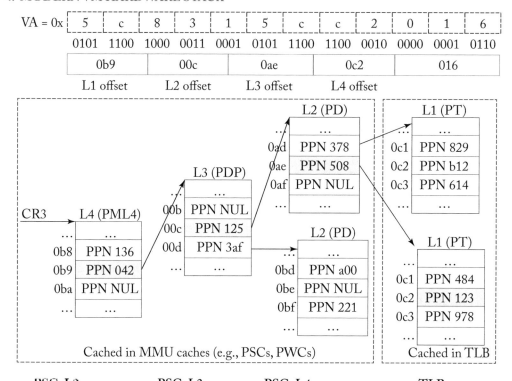

Figure 4.7: x86-64 page table walk for virtual address 0x5c8315cc2016. TLBs cache L1 PTEs and MMU caches store L2-L4 PTEs. Conventional caches can store all entries [18].

TLB miss handling in the UltraSPARC family of CPUs. Since UltraSPARC CPUs implement TLB miss handling via a trap mechanism, performance of the low-level trap handling code in the operating system is crucial to overall system performance.

Like other software data structures, TSBs are cached in the conventional on-chip hardware caches. A special CPU register maintains a pointer to the physical address storing the root of the TSB. After this register is used to look up the TSB, the requested virtual page number is used to index into the desired location. The TSB is maintained as a direct-mapped structure, with a practically unbounded capacity. Overall, TSBs are used as follows. On a TLB miss, an interrupt runs the OS code. This OS handler then searches the TSB for a valid entry whose tag matches the virtual address of the translation miss. If the entry is found, it is loaded from the

TSB into the TLB, and the trapped instruction is replayed. If the entry is not found, the page table is then walked.

Prior to Solaris 10, the user process TSBs used to come from a global pool of fixed size which was allocated at boot. In Solaris 10, Solaris began implementing dynamically allocated TSBs. While the switch to hardware-managed TLBs has largely supplanted the use of TSBs, emerging hardware accelerators with potentially alternate page table organizations and TLB management strategies may revive the befits of TSBs.

4.7   SUMMARY

In this chapter, we covered the hardware and ISA-level design space of the VM subsystem. We explored alternative page table designs, and we discussed how TLBs and other caching structures are spread throughout an architecture to accelerate translations and to keep the VM subsystem off the execution critical path as much as possible. With this, we now jump upwards in the computing stack into a discussion of the software layers of the VM subsystem.

CHAPTER 5

Modern VM Software Stack

Having presented details on VM hardware, we now turn our attention to OS-level support. In general, the OS contribution to memory can be divided into two components. First, low-level OS code must be tailored to match the details of the VM hardware (e.g., TLBs, page table walkers, etc.). Second are higher-level software operations that are somewhat abstracted from the low-level hardware details. In this chapter, we focus on the higher-level decisions that the OS must make in order to manage the VM subsystem efficiently.

Before diving into the details, we first recap the various levels of abstraction at which memory needs to be allocated. The allocation of VM blocks within a process' virtual address space is a concern of the operating system. The OS must attempt to space out allocated memory regions across the virtual address space such dynamically sized regions do not collide and such that fragmentation does not prevent future regions from being created. With 64-bit systems and virtual address spaces with 48–52 bits (and room to expand), this is much less of a concern than it is with 32-bit systems (or narrower). The allocation of objects within a single block of VM and the responsibility to deal with fragmentation are concerns for user-level code, libraries, and runtime systems. While these allocators can also be very important for performance, they are largely transparent to the VM subsystem, and we do not discuss them further here.

The challenge we spend more time focusing on is page frame allocation: the assignment of physical page frames to each VM page. The physical memory space is often capacity-constrained and/or fragmented on today's systems [74, 90, 91], and so page frame allocation becomes an important responsibility for the OS to manage. Emerging hardware optimizations such as TLB coalescing also require the OS to make intelligent page frame allocation decisions [32, 90, 111]. In the rest of this chapter we therefore dive into the details of page frame allocation as it is performed by today's operating systems.

5.1 VIRTUAL MEMORY MANAGEMENT

One of the most important data structures maintained by the OS is the one that tracks the VM regions associated with each process. Linux, for example, uses a VM area (VMA) tree made up of VMA areas [30, 31]. Pointers to the VMA tree are maintained by Linux's per-process `mm_struct` data structure or main memory descriptor. There is one main memory descriptor per address space. The VMA tree records all the VMA regions used by the process. Each VMA region is a contiguous range of virtual addresses, which never overlap. Furthermore, the size of each VMA is a multiple of the page size of the system.

Linux tracks two types of VMA mappings in its VMA tree: (1) file-backed mappings allocated using mmap() system calls, which are used to represent code pages, libraries, data files, and devices; and (2) anonymous mappings, which represent regions such as the stack and the heap not backed by a file. For each of these mappings, VMA regions maintain a pointer to the start and end virtual address of each region and page protection bits that indicate whether the page is readable, writeable, or executable. Each VMA region also maintains VMA protection bits or flags, which are a superset of the page protection bits. Figure 5.1 shows an example of mmap'ed VMA regions.

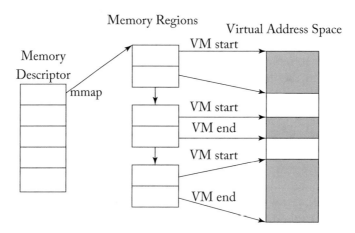

Figure 5.1: The memory descriptor points to a region mmapp'ed with multiple memory regions or VMAs. These point to the start and ends of various contiguous blocks of the virtual address space.

Through the course of a program's execution, VM regions are added to and deleted from existing VMAs. Figures 5.2 and 5.3 shows how existing VMA regions are enlarged and shortened on these operations. In general, VMA areas are enlarged whenever an new file is mmap'ed, a new shared memory segment is created, or a new section is created (e.g., for a library page, code page, heap page, or stack page). The kernel tries to merge these new pages with existing adjacent VMAs.

Because of the critical nature of its operation, the VMA tree is a frequently accessed data structure. Every page fault, every mapping operation, etc., requires a VMA tree lookup. Since a process may have several VMAs, the VMA tree data structure must enable quick lookup. In Linux, as with many other data structures, VMA trees are implemented with red-black trees because they facilitate an O(log(n)) search time [50].

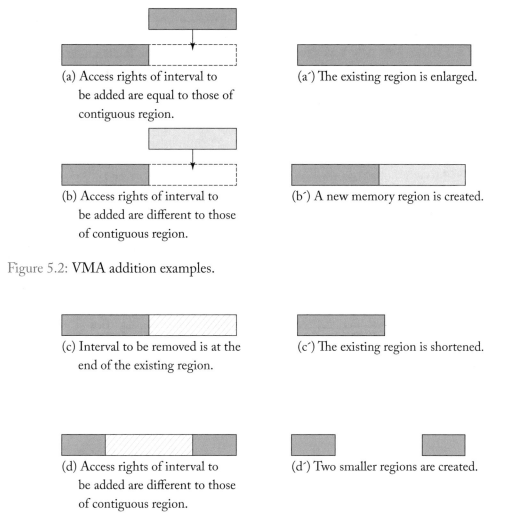

(a) Access rights of interval to be added are equal to those of contiguous region.

(a′) The existing region is enlarged.

(b) Access rights of interval to be added are different to those of contiguous region.

(b′) A new memory region is created.

Figure 5.2: VMA addition examples.

(c) Interval to be removed is at the end of the existing region.

(c′) The existing region is shortened.

(d) Access rights of interval to be added are different to those of contiguous region.

(d′) Two smaller regions are created.

Figure 5.3: VMA removal examples.

5.1.1 DEMAND PAGING AND LAZY ALLOCATION

There are two approaches to managing new VM allocations. In one approach, all the additional virtual pages added to the VMA could be immediately assigned new physical page frames and the page tables could be immediately changed to reflect these assignments. The problem with this approach is that it wastes memory, since it is not yet known whether the process will actually access all the newly allocated virtual pages [88, 90]. Therefore, most OSes use a different approach, called lazy allocation, a form of demand paging. In this approach, physical page frames are not immediately assigned to the new virtual pages. Instead, allocation occurs only when the program tries to access each new virtual page for the first time.

Figure 5.4 shows how new VM allocations are handled on page faults under Linux. In step 1, a program uses a `malloc()` call, which eventually (at least in some implementations of `malloc`) makes an `brk/sbrk()` system call to grow its heap. In step 2, `brk()` enlarges the heap VMA. In step 3, the processor attempts to access the new page for the first time. When it attempts to translate the accessed virtual address to a physical address, it encounters an unassigned translation in the TLB and page table. This prompts a page fault, invoking the OS. Anonymous pages do not require data transfer from disk, so they result in low-latency minor page faults, while file-backed page require data transfer from disk and hence trigger major page faults. In our example, the growth of the heap is a minor page fault. At this point, the OS sees that the virtual page is being requested, and therefore assigns a physical page to it, creating a page table entry (shown in step 4). Naturally, this description is a simplification of the actual page fault handling code in Linux. We refer readers to the Linux code for more details.

Program calls brk() to grow its heap

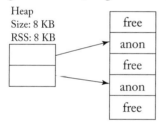

brk() enlarges heap VMA; new pages are not mapped onto physical memory

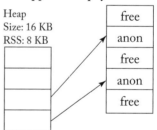

Program accesses new memory CPU page faults

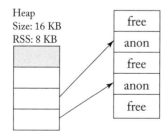

Kernel assigns page frame to process, creates PTE, resumes execution.

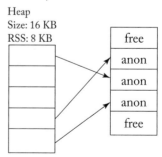

Figure 5.4: Dynamic memory management and interactions with page faults.

5.1.2   COPY-ON-WRITE

Similarly to how a page frame is not allocated as soon as a new virtual page is created, the same is true when a page is duplicated. Copy-on-write (CoW) is a clever scheme by which a memory range that is duplicated into a new region is not physically allocated a new page frame until some data in one of the two pages in question is actually modified [91, 98]. Pages may be duplicated for any number of reasons. For example, a process may fork, thereby requiring the operating system to create a duplicate of the process' entire virtual address space. In these cases, duplicating the physical page frame as soon as the virtual page is duplicated would be a waste of resources. For one thing, forked processes often quickly discard their inherited address space in order to begin executing the code of another program, meaning that any duplicated page frames would be quickly thrown away. In many other situations, the pages are read-only, so that there is no need to duplicate page frames there either.

Instead, copy-on-write implements the page frame duplication lazily. At first, the original page frame remains, and the two virtual pages each point to it. Both are marked read-only so that any attempts to write to the page will trap into the operating system. When such a write happens, only then does the page frame get duplicated, and each virtual page is reassigned to point to its own copy of the page (with the full original permissions restored). From then on, the pages and page frames behave as normal independent entities.

Copy-on-write is implemented by manipulating the page permission bits. When an OS wants to mark a page as copy-on-write, it marks the page table entry as non-writeable, but in its own internal data structures it tracks the VM region associated with the page as writable. Therefore, when the processor executes a store instruction to the memory address, the TLB (or page table) triggers a protection fault, since writes are not permitted. At this point, the page fault handler looks up the OS data structures and discovers that the page is marked copy-on-write. Consequently, the page fault handler makes a duplicate of the physical page, updates the page table entry, and returns execution to the program.

5.1.3   ADDRESS SPACE LAYOUT RANDOMIZATION

Having detailed the key data structures involved in managing VM, we now discuss some additional enhancements used to enable better memory security. Specifically, Address Space Layout Randomization (ASLR) is VM technique used to protect against buffer overflow and other types of security attacks, and has largely been adopted by most mainstream OSes since 2001–2003 [47, 71, 100].

ASLR hinders attacks by making it more difficult for attackers to predict process target addresses. It does this by randomly arranging the address space positions of key data areas including the base of the executable, and the positions of the stack, heap, and libraries. This makes it difficult, for example, for attackers trying to execute return-to-libc attacks to locate the code to be executed, or for attackers trying to inject shellcode on the stack to identify the stack position. In such cases, the system obscures memory addresses from the attackers. This means that these memory addresses have to be guessed, and a mistaken guess is usually not recoverable since the application crashes.

Since ASLR hinges on the low probability of an attacker guessing the locations of randomly placed areas, security is increased by making the search space larger. Therefore, ASLR is more effective when there is more entropy present in the virtual address space. Entropy can generally be increased by increasing the number of VMA regions over which randomization occurs.

Linux enabled a weak form of ASLR in 2005. Subsequent patches called PaX and Exec Shield enabled better ASLR, and is used by default by various Linux distributions including Alpine Linux, Hardened Gentoo, Hardened Linux from Scratch, etc. Beyond PaX, today Linux also allows for position-independent executable (PIE), which implements a random base address for the main executable binary. PIE essentially ensures that the base executable address is randomized as effectively as the shared libraries. In tandem with PIE, modern Linux distributions also use kernel address space layout randomization (KASLR), which brings ASLR support for the kernel pages themselves by randomizing where the kernel code is placed at boot time [41, 60]. KASLR has been merged into mainline Linux since 2014.

Since Vista's release in 2007, Windows OSes have ASLR enabled for executables and dynamically linked libraries specifically linked with ASLR-on flags. Other processes do not have ASLR enabled by default, to ensure backward compatibility. Turning ASLR on randomizes the location of the heap, stack, process environment block, and thread environment block. Finally, Apple introduced ASLR for system libraries in 2007. In 2011, Apple expanded ASLR implementation to also cover all applications for 32-bit and 64-bit implementations. Since 2012, Apple has integrated kernel-level ASLR as well.

5.2   MANAGING LOCALITY

When physical memory capacity is constrained, then some pages must be swapped to disk in order to make room for others. One key technical question is how to identify the important portions of an application's memory footprint, to ensure that the most useful pages are the ones actually preserved in memory. To this end, VM uses the notion of a working set.

5.2.1   WORKING SETS

The working set of a program is one of the most important concepts in computer science and defines the amount of memory that a process requires in a given time interval. Naturally, this is the least amount of process memory that a VM subsystem should aim to accommodate in physical memory at any given time. The notion of a working set was posited by Peter Denning in his classic papers [34–36]. The working set $W(t,a)$ of a process at time t is the collection of information referenced by the process during the process time interval $(t - a, t)$. With paging-based VM, the units of information in question are memory pages. Due to program locality behavior, the theory of working sets anticipates that the set of pages that the process will access in in time (t, t + a) can be approximated by the pages accessed during (t − a, t).

The seemingly straightforward notion of working sets has important implications on systems design. If too many pages of a process are kept in a capacity-constrained main memory, then fewer other processes can be ready at any one time. If too few pages of a process are in memory, the page fault frequency is increased, and processes become mostly inactive or suspended waiting for disk transfer. Therefore, the working set of a process must be approximated with care, so that it is a reliable predictor of future program behavior.

The key mechanism used to identify the working set of a program is to distinguish between memory pages that have been referenced by a program, and those that have not. The referenced pages are kept in main memory, while those that have not been referenced in a long time are candidates for eviction from main memory (when pages need to be brought in from disk). To enable VM, we need mechanisms to achieve the following three objectives: (1) detect references to pages on memory or disk, to identify a program's working set; (2) choose what pages in main memory to evict to make room for incoming pages from disk; and (3) choose when to bring in pages from disk to main memory. We now present approaches to solving these three problems.

5.2.2 NAIVE PAGE REPLACEMENT POLICIES

The goal of a page replacement policy is to make the most efficient use of the accessed bits. We describe various schemes below.

We first discuss the optimal scheme. Although not practically realizable, it presents a useful counterpoint to practical algorithms. Past work has shown that the optimal algorithm, also called Belady's algorithm, evicts the memory page that will not be used for the longest time [14]. That is, the optimal algorithm relies on oracle knowledge of future access patterns and evicts the page that is references furthest ahead in the future. Figure 5.5 shows such an optimal algorithm for a set of successive memory references to physical pages (1, 2, 3, 4, 1, 2, 5, 1, 2, 3, 4, 5). We show the contents of physical memory, which can accommodate 4 page frames, for every reference, and we find that the minimum number of page faults for our references is 6.

At the other end of the spectrum, we can consider naive schemes such as random or FIFO replacement. The advantage of random replacement is its simplicity. With a random policy, there is no need to maintain information about which pages were referenced in the past. The downside of this approach is of course sub-optimal replacement policy decisions. Since the performance impact from poor page replacement is harmful to performance, random replacement policies are rarely used in practice.

First-in first-out (FIFO) algorithms achieve a compromise between simplicity of implementation and good choice of page replacement candidates [106]. The kernel keeps track of the order in which pages were brought into main memory, usually by maintaining a kernel-level linked list of in-memory pages, where the head represents the most recently allocated page. On page eviction, the oldest page (from the tail of the list) becomes the prime replacement can-

References	1	2	3	4	1	2	5	1	2	3	4	5	
	1	1	1	1	1	1	1	1	1	1	4	4	
Replacement Policy Ordering	⊘	2	2	2	2	2	2	2	2	2	2	2	6 Page Faults!
	⊘	⊘	3	3	3	3	3	3	3	3	3	3	
	⊘	⊘	⊘	4	4	4	5	5	5	5	5	5	

Figure 5.5: An optimal page replacement algorithm evicts the page used furthest ahead in the future.

didate. The intuition for FIFO policies is that pages that were allocated the furthest back in time are also possibly the least likely to be used in the future. However, in practical scenarios, the performance of FIFO is generally below that of more sophisticated schemes and so it is not commonly used. Figure 5.6 shows the impact of FIFO replacement on the memory accesses previously discussed for optimal replacement. Note how in the same example as above, FIFO suffers 10 page faults vs. the 6 of optimal replacement.

References	1	2	3	4	1	2	5	1	2	3	4	5	
	1	1	1	1	1	1	5	5	5	5	4	4	
Replacement Policy Ordering	⊘	2	2	2	2	2	2	1	1	1	1	5	10 Page Faults!
	⊘	⊘	3	3	3	3	3	3	2	2	2	2	
	⊘	⊘	⊘	4	4	4	5	5	5	5	3	3	

Figure 5.6: FIFO replacement evicts the page brought furthest back in time.

5.2.3   LRU PAGE REPLACEMENT POLICIES

By far, the most commonly used replacement algorithm tracks the LRU memory pages, and prioritizes them as eviction candidates. LRU is considered to be a reasonably good policy for most workloads. The primary problem with LRU is implementing it efficiently. Direct implementation of LRU will often maintain counters or timestamps with every page [35]. These counters are updated on memory accesses, and on page replacement, all counters are scanned to identify the oldest page. While conceptually implementing strict LRU, linear scans are expensive while maintaining full timestamps consumes non-trivial memory. Alternately, software approaches replace the need for counters by maintaining a linked list of pages. Whenever a page is referenced (or when it is brought into memory for the first time), it is added to the head of the head of the list. Pages at the tail of the list are therefore those that have been referenced furthest back in time and are the prime candidates for eviction.

Figure 5.7 shows the operation of an ideal LRU algorithm. While LRU cannot achieve the optimal algorithms 6 page faults, it does outperform FIFO, which suffers 10 page faults. While such pointer-based LRU implementations eliminate the linear scan times of hardware approaches, they require complex pointer management, precluding its use. Therefore, most modern OSes (including Linux, FreeBSD, and Solaris) use approximate LRU (or pseudo-LRU) algorithms [4, 27, 35]. Approximate LRU algorithms can enable much simpler implementations while not giving up much in the way of performance, making them a common choice of policy.

References	1	2	3	4	1	2	5	1	2	3	4	5	
	1	1	1	1	1	1	1	1	1	1	1	5	
Replacement Policy Ordering	⊘	2	2	2	2	2	2	2	2	2	2	2	8 Page Faults!
	⊘	⊘	3	3	3	3	5	5	5	5	4	4	
	⊘	⊘	⊘	4	4	4	4	4	4	3	3	3	

Figure 5.7: LRU replacement evicts the page used furthest back in time.

The most widely adopted approximate LRU algorithm is the CLOCK algorithm [27]. CLOCK harnesses the page table entry accessed bits for each page. As previously described, these bits are set by the page table walker on a TLB miss, when a page table lookup completes and an entry is filled into the TLB. Periodically, the OS traverses the page tables, setting all accessed bits to zero to ensure that page reuse information does not become stale. As the pages are accessed over time, their accessed bits are set once again. At any given point in time, the set of pages whose accessed bits are set therefore represents not just the set of pages that have ever been accessed, but rather the set of pages which have been accessed recently, i.e., since the last clearing of the bits. In this way, the accessed bits can be used to track some approximation of the working set of a process.

There are several ways to implement the CLOCK algorithm in practice. One of the best-known approaches is to combine the software implementability of FIFO with the hardware efficiency of CLOCK in an approach called LRU with second chance [39]. In this approach, all pages are conceptually maintained in a circular list. A pointer (called the clock hand) is maintained to the next page replacement victim. When a page has to be replaced, the OS examines the page that the hand points to. If the accessed bit for this page is set, the hand is advanced, and the accessed bit cleared. Otherwise, the current page is chosen as the victim.

Figure 5.8 shows an example of the CLOCK algorithm's execution. The string of virtual page numbers being accessed is shown on the top. Underneath each reference, we present the memory state, comprised of 4 physical pages. For each physical page, we show the virtual page mapped to it, and whether its accessed bit is set. Therefore, (1/1) indicates virtual page 1 with its accessed bit set, while (2/0) indicates virtual page 2 with its accessed bit cleared. As before, highlighted page/accessed bit combinations indicate the recently allocated virtual page. Further, the arrow presents the CLOCK hand. The first four memory references map the four physical frames to virtual pages 1–4, setting all their accessed bits. The CLOCK hand remains at 1/1 because there is no need to evict a page up to this point. However, on the first reference to 5, a page must be replaced. The CLOCK algorithm hand moves down the list of physical frames, checking for one with a clear accessed bit. Since none of the physical frames have a clear accessed bit, the hand moves back up to the first frame, after clearing all the accessed bits it encounters. It then evicts the contents of virtual page 1, and assigns it to virtual page 5 (setting its accessed bit). The hand now points to the second physical frame. Therefore, when virtual page 1 is next accessed, the CLOCK algorithm checks the physical frame pointed to by the hand. Since the accessed bit is clear, this is a candidate for page replacement. Therefore, the contents of virtual page 2 are evicted, and virtual page 1 is assigned the second physical frame. Figure 5.8 shows the remainder of the example.

Refs.	1	2	3	4	1	2	5	1	2	3	4	5
	→1/1	→1/1	→1/1	→1/1	→1/1	→1/1	5/1	5/1	5/1	→5/1	4/1	4/1
Rep. Policy Ordering	⊘	2/1	2/1	2/1	2/1	2/1	→2/0	1/1	1/1	1/1	→1/0	5/1
	⊘	⊘	3/1	3/1	3/1	3/1	3/0	→3/0	2/1	2/1	2/0	→2/0
	⊘	⊘	⊘	4/1	4/1	4/1	4/0	4/0	→4/0	3/1	3/0	3/0

10 Page Faults!

Figure 5.8: Example of the CLOCK replacement algorithm.

Most CLOCK algorithms differentiate between clean pages (whose data does not have to be written back to disk) and dirty pages (whose disk copy has stale data). Since dirty pages need to be written back to disk, they are more expensive to evict. As a result, the CLOCK algorithm first tries to evict a page whose access and dirty bits are cleared. If no such page exists, it chooses a page whose accessed bit is cleared but dirty bit is set, writing back the page's data to disk. A special kernel daemon is responsible for performing the page write-back.

5.2.4   PAGE BUFFERING

Eviction is not the only operation that the OS tries to optimize with its paging decisions. I/O traffic (e.g., to disk) can be very expensive, and so the OS will generally implement intelligent buffering schemes to try to take the penalty of performing I/O off of the critical path [62–64, 118]. Sometimes, this means being aggressive with write-back of dirty data to a page. If a dirty page seems likely to be evicted soon, then it can make sense to perform a write-back ahead of the actual eviction, thereby removing the write-back itself from the critical path of evicting that page. This is achieved by implementing intelligent page buffering. The general idea is to always keep a pool of free physical page frames. When the number of pages in this pool decreases beyond a preset threshold, the OS preemptively writes back and/or evicts pages according to the replacement algorithm, adding them to the pool of free frames. Pages can be evicted when convenient; for example, the OS may choose to write pages back when there is little disk traffic. Once this is done, the page can be marked clean in the page table.

It is important to note that writing pages back to the disk does not mean that physical frames in the free buffer lose their contents. If the page is reused in the near future, it can again be mapped from the free pool quickly as long as the data is still physically present, obviating the need for expensive disk transfers to bring the data back in. If, on the other hand, the frame is needed to make room for a demanded and incoming disk page, the page can be deleted without the need for the write-back of the dirty data to disk to be on the program's critical path.

Other times, being intelligent about I/O means *not* writing back content to disk right away, but instead allocating a page cache to buffer I/O accesses. These buffers intercept all accesses to the I/O memory region in question, and because the buffers themselves live in memory rather than in disk, they can be accessed much more quickly. Then, whenever the page is evicted, and/or whenever the device is being shut down or ejected from the system (e.g., when removing a USB device), the CPU caches and page caches are flushed and written back to the non-volatile I/O storage.

5.3   PHYSICAL MEMORY ALLOCATION

The primary challenge in designing memory allocators is to manage fragmentation effectively [15, 16, 114]. Systems suffer from two types of memory fragmentation. External fragmentation is a form of memory fragmentation that is visible to the allocation system. This refers to "holes" in memory that are not reusable because any single one is smaller than the size of a new memory region being allocated. In contrast, internal fragmentation is visible to just the process and refers to the amount of wasted space within a single allocation unit. For example, if a desired unit of allocation cannot be accommodated within the free space inside of an existing page, then the unusable space is wasted. The goal of a dynamic memory allocator is hence to reduce the number of holes in memory, and to keep the holes sufficiently large so that both types of fragmentation are reduced [15].

5.3.1   NAIVE MEMORY ALLOCATORS

We start by discussing the drawbacks of some naive memory allocators. One straightforward approach is known as best fit allocation [15, 73]. In this approach, the entire free list (i.e., the list of free memory regions) is searched on each memory allocation request. The allocator then chooses the smallest block that can satisfy the request. (Alternatively, if an exact size match is found, the search can be stopped early.) There are several problems with this approach. First, it involves searching most or all of the free memory to identify the best fit. While search time could be reduced by designing the free list data structure intelligently, it remains high overhead. Second, best fit allocation schemes tend to leave very large and very small holes in the memory address space. The small holes in particular become hard to use as they are usually too small for future allocation requests. Figure 5.9 shows an example of the best fit algorithm. We assume two free chunks of memory, of 20 pages and 15 pages in size. Further, we assume that there are two requests for 10 virtual pages, and 20 virtual pages. Both these requests can be accommodated in this example, but a small hole is left in the chunk of 15 pages. Figure 5.10 shows the problems with such holes, for a different set of requests for 8, 12, and 13 pages. In this example, the request for 13 pages fails.

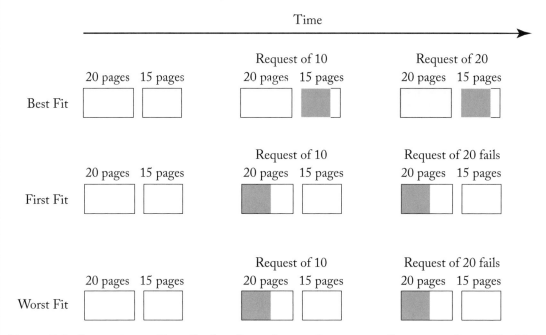

Figure 5.9: Comparison of best fit, first fit, and worst fit memory allocation policies. The blue chunk of memory shows an allocation for 10 pages, and the green for an allocation of 20 pages.

First fit approaches are quicker than best fit as they allocate the first free space of memory with sufficient capacity for the allocation request [73]. The problem with this approach is that

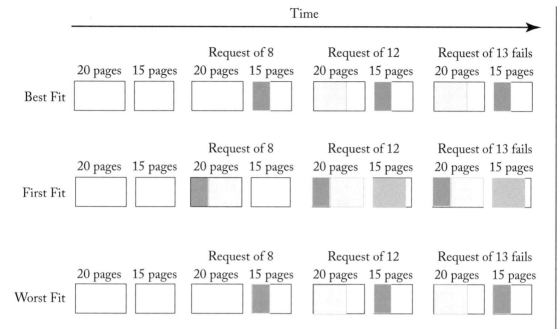

Figure 5.10: Comparison of best fit, first fit, and worst fit memory allocation policies. The blue chunk of memory shows an allocation for 8 pages, the green for an allocation of 12 pages, and the gray for an allocation of 13 pages.

there are generally even more holes left unused in the heap than with best fit algorithms. Figure 5.9 shows an example of a first fit approach for a sequence of memory allocation requests. As shown, the second request for 20 pages fails because the first request uses up too much of the first contiguous chunk of free memory.

5.3.2 BUDDY ALLOCATION

First and best fit approaches suffer from non-trivial drawbacks. Therefore, modern systems implement a better data structure to track free blocks. The idea of this approach, used by OSes like Linux, FreeBSD, etc., is to use a buddy allocator [72]. The buddy allocator is designed on some key observations. First, memory allocation requests can be effectively pre-sorted by size into different categories. This insight can be used to overcome the lookup problems of generic allocators like first and best fit. Second, the data structure must not only allocate quickly but also free memory quickly. This means that the memory allocator must avoid iterating through the entire free list to achieve these aims. Third, unsurprisingly, it is beneficial to maintain free physical pages as contiguously as possible in order to minimize external fragmentation.

Figure 5.11 shows how a buddy allocator is designed. First, units of allocation are restricted to byte sizes in powers of 2. The buddy allocator maintains a number of lists, numbered from

0 to K. List 0, 1, 2, and K maintain information about 2^0, 2^1, 2^2, and 2^K free contiguous bytes respectively. Suppose there is a memory allocation request for 2^N bytes. The buddy allocator applies a ceiling function to N, to find the smallest power of 2 value larger than N. The list for N is first searched (followed by the list for N + 1, N + 2, etc.) until a free chunk of 2^N bytes is found.

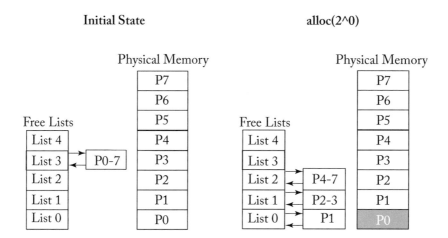

Figure 5.11: Buddy allocator lists maintain information about power-of-2 contiguous free pages. Allocated pages are shown in blue. The figure on the left shows 8 unallocated page frames. List 3 in the buddy list maintains information about 8 runs of contiguous free pages. The figure on the right shows the state of physical memory and the buddy allocator after a request for 1 page. Allocated physical pages are shown in blue.

If the only free blocks available in a buddy allocator search are bigger than 2^N, the free block is recursively halved until a region of the appropriate size becomes available. That block is then allocated, while the other recursively split buddy blocks are inserted into the appropriate list positions. As a result, the buddy allocator cleverly continues to track a set of available blocks with as much contiguity as possible under the powers-of-2 scheme. Buddy allocator deallocation or free operations are also straightforward. Whenever a memory block is freed, it is merged with any contiguous free blocks with which it is properly aligned. The merged block is then inserted into the appropriate buddy position.

Figure 5.11 shows how the buddy allocator changes for an allocation of 1 page. In our example, 8 frames of physical memory are initially unallocated. Since these frames are contiguous, list 3 is used to record them. When an allocation of 1 page is made, the buddy list is recursively split so that there is one entry for a contiguous list of 4 pages, a contiguous list of 2 pages, and a singleton. Naturally, 1 page (P0) is allocated. The diagram on the left in Figure 5.12 then shows how an allocation of 4 pages changes the buddy list. The list of 4 contiguous pages (the entry in list 3) is used up. Figure 5.12 also shows how freed pages can result in the re-formation of a block of larger contiguity.

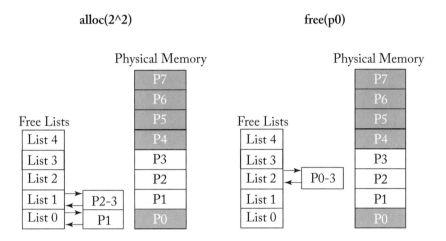

Figure 5.12: The figure on the left how an allocation for 4 pages is performed, starting again from the example in Figure 5.11. The figure on the right shows the state of the buddy allocator after P0 is freed.

5.3.3   MEMORY POOLS AND SLAB ALLOCATION

Buddy allocators are better than generic allocators in several ways [15, 72, 73]. First, because they organize free blocks in lists associated with their size, they enable faster allocation and deallocation. Second, by merging free physical blocks, they reduce fragmentation. Their main problem, however, is their inability to eliminate internal fragmentation. Since allocations occur in power-of-2 sized blocks, there may be unused space at the end of these blocks. The smaller these unused spaces, the harder it is to allocate them eventually. User libraries and/or code expecting to allocate objects at sizes not close to a power of 2 will often therefore establish memory pools within larger blocks of VM, and then provide a separate (user-level) allocator for objects placed in that block.

Most OSes also maintain special slab allocators dedicated to kernel-level memory allocations [23]. Slab allocators are motivated similarly: kernel allocation requests are often for particular objects (e.g., inodes) with well-known size requirements. Slab allocators handle these requests more efficiently than general-purpose allocators. Slab allocators are made up of several software "caches" of slabs. These slabs maintain memory slots for different types of kernel objects like inodes, task control blocks, etc. A cache may contain several slabs, each of which is a physical page in size. Each slab maintains a bitmap which records whether its slots are allocated or free. A request for a specific object is routed to the appropriate cache of slabs. The bitmap is used to identify a target slot. The relevant bitmap entry is then set and the slab slot is returned to the kernel for allocation. A deallocation of the object merely requires the relevant bitmap entry to be cleared [23].

5.3.4   PAGE COLORING

As we saw in Section 4.2.2, page coloring refers to a scheme in which a memory allocator "colors" each page according to the position (i.e., the set) into which it will be placed in the cache. Section 4.2.2 discussed the use of page coloring for correctness, but since most modern VIPT L1 caches follow the associativity guidelines discussed in that chapter, page coloring is not generally needed for that reason.

Page coloring is instead often considered as a mechanism for improving performance with caches that are not fully associative. Suppose in a pessimistic case that every page frame in a process' working set ends up mapped into the same set of the cache. In such a case, that set of the cache will be thrashed heavily (i.e., it will face a lot of extra misses due to the data not fitting into the set), but the other sets will sit idle and underutilized! To avoid this, various authors have proposed to use page coloring to ensure that all of the pages in a working set and/or in each allocation are distributed across different pages as much as possible (but still subject to synonym constraints where applicable!) [69]. In practice, the benefits of page coloring for performance (along with high-associativity caches) have not been universally shown to outweight the added cost placed onto the critical path of memory accesses, and so page coloring is not a universal solution today. Nevertheless, researchers continue to explore new ways in which page coloring can be improved [116].

5.3.5   REVERSE MAPPINGS

We discuss one final memory-tracking data structure: the reverse mapping. Reverse mappings are necessary to address the following problem: the VM page replacement algorithm may opt to swap out a physical page mapped into the virtual address space of one or more processes. In other words, multiple page table entries may point to the same physical page. Unfortunately, when the page replacement algorithm opts to swap out a physical page, it only knows the identity of the physical page, and not the virtual pages and page tables pointing to it. Reverse mappings are necessary to identify these virtual pages and their page table entries, both for single-user pages and for synonyms.

Reverse mapping data structures have a design space very similar to that of the page table, with many of the same kinds of tradeoffs. They are not as critical to performance, since they are less common. However, they are still often implemented in the form of priority search trees so that queries can nevertheless be made as efficient as possible [37].

5.4 SUMMARY

In this chapter, we discussed some of the functionality and tradeoffs present in the operating system's portion of the VM subsystem. The OS is responsible for managing the layout of the virtual address space and the physical address space of each process. It is also responsible for balancing resource requirements across multiple processes when capacity is constrained and/or fragmented. As such, modern OSes have developed a number of clever tricks such as copy-on-write to ensure that resources are not unnecessarily wasted and that the VM management bookkeeping procedures are kept as efficient as possible.

From here, we shift gears away from exploring design spaces and into a discussion of some of the more advanced use cases and subtle correctness issues in the world of VM. In particular, we start by exploring parallelism, and then we move on to a study of how the VM subsystem is kept synchronized in parallel systems.

CHAPTER 6

Virtual Memory, Coherence, and Consistency

In previous chapters, we discussed how page table updates take place. The operating system stores new values to the page table in memory, and then the hardware or software page table walker performs loads to memory and/or to specialized caches in order to bring those newly updated page table entries into the TLB for subsequent use during normal memory accesses. At a very basic level, these VM subsystem update mechanisms are just loads and stores. The practical reality, however, is much more complicated.

The fundamental challenge that we discuss in this chapter is the fact that TLBs are not generally kept *coherent* with the rest of the memory system: stores to memory to update the page table do not automatically propagate to the TLBs, nor are stale TLB entries even automatically invalidated! Most general-purpose processors provide a hardware cache coherence protocol to ensure that data caches are kept coherent, but the same is not generally true for TLBs. In fact, instruction caches are not always kept coherent either. This lack of coherence places the burden on the programmer and on the OS to add explicit synchronization whenever the state of the VM subsystem is updated.

In this chapter, we take a closer look at the reasoning behind the lack of hardware coherence for TLBs and (often) instruction caches. We then study the synchronization requirements that this lack of hardware coherence imposes onto programmers, for single-threaded and for multi-threaded code. At the end we also briefly discuss how even caches that are kept coherent can produce unexpected behavior in the form of a weak memory consistency model.

6.1 NON-COHERENT CACHES AND TLBS

Although most CPUs dedicate hardware resources to keeping caches coherent, they do not generally do the same for TLBs (nor, often, for instruction caches either). Like everything else with architecture, it all comes down to understanding the tradeoffs between the performance, power, area cost, and programmability/ease of use of each alternative design point. Data caches hold data that is read and written frequently, sometimes by more than one core at a time. The added complexity of requiring software-managed coherence for data caches have generally been considered too large of a burden for programmers, and therefore architects are willing to dedicate extra transistors to build a cache coherence protocol that keeps data caches coherent automatically.

Before we study how TLBs are non-coherent, let us first recall how coherence is defined and implemented. A memory system is defined to be coherent if it provides a globally agreed-upon total order of the stores to each memory location, if each load returns the value written by the latest store to the same memory location, and if each store (or a coherence successor thereof) propagates to every other observer in a finite amount of time. The first requirement is sometimes restated as a single-writer/multiple-readers (SWMR) condition: at any given point in time, there can either be a single entity with write access to a memory location, or any number of entities with read-only access to the memory location. This condition imposes a total order on the writes to each memory location, but it also adds conditions to reads as well, making it a slightly stronger requirement. The second condition for coherence ensures that loads actually return the expected values. The major complexity here is in defining the meaning of "latest;" this can only be done properly via the memory consistency model (Section 6.4.2). The third condition simply ensures that a processor will always make forward progress when executing multithreaded programs.

Cache coherence protocols satisfy the coherence requirements above by adding bits in hardware to track the state of each cache line. The cache tracks whether each cache line is in a *shared* state (i.e., with read-only permission), a *modified* state (i.e., with read-write permission and exclusive ownership), or an *invalid* state. (Real coherence protocols of course have many more states than we describe here [101].) Whenever a core wants to do a store into a line in invalid or shared state, it must first explicitly invalidate all other caches holding the same line, so that the cache that sent the invalidations can legally move its line into modified state. In this way, every cache in the system is quickly and directly notified whenever a core writes to a memory location being accessed by more than one core. Likewise, whenever a core wants to do a load from a line currently in the invalid state (or not present), it queries the coherence subsystem (e.g., the directory) to find and fetch the latest data for that memory location.

On the other hand, TLBs and instruction caches store data that is mostly read-only and which changes relatively infrequently, if at all. Furthermore, instruction fetches and page table walks often take place through different pathways in the microarchitecture from normal memory loads. Extending a coherence protocol into all of these other caches and TLBs would therefore add non-trivial area and power overhead for much less of a clear benefit that is seen with data caches that see more frequent writes. As a result, hardware coherence is generally considered unnecessary for TLBs and instruction caches.

In fact, it would be hard to make TLBs coherent even independent of the performance/power/area tradeoffs. Because TLBs are not kept hardware-coherent, TLB entries are not generally tagged with the physical address from which they were originally fetched. This means it would be impossible to successfully snoop on the coherence protocol traffic even if the messages were routed to the TLB! It would be possible to augment TLBs to include this information, but even then the mechanism is non-obvious. The authors of UNITD observed that a set-associative TLB is indexed by bits of the virtual address being looked up, not by bits of the physical address where the page table entry actually lives [95]. This means that if a TLB wanted to participate in (or at least snoop on) the cache coherence protocol, it would have to search the entire TLB to find entries matching the tag of the message, rather than needing just to search a particular set. Therefore, the UNITD authors proposed to add a fully associative Page Table Entry CAM (PCAM) to track the mapping between TLB entries and page table entries, and they demonstrated how the PCAM could enable the TLBs to participate successfully in a cache coherence protocol.

Unless ideas such as UNITD are implemented, the lack of a coherence protocol for TLBs and/or instruction caches means that update notifications are not propagated automatically to each TLB and/or instruction cache. In practice, most TLB entries will eventually be evicted naturally due to a context switch, due to replacement of another line in the same set, or due to any number of other reasons that come up naturally during execution. However, a stale TLB entry can be dangerous if there is even a small finite amount of time during which it can be used. It therefore falls to software to ensure that the VM subsystem is synchronized in such a way that stale TLB entries are in fact invalidated before they can cause any problems. We discuss these procedures in the coming sections.

6.2 TLB SHOOTDOWNS

Before we discuss how TLBs are synchronized with page table updates, let us first consider what can go wrong if this synchronization is not performed. Suppose a page frame belonging to some multithreaded process is being swapped out to disk. The OS will copy the data to the backing store, and then it will invalidate the page table entry (or entries) in the page table pointing to that physical address range, using the procedure described in Section 3.4. At this point, the page table itself is updated, and subsequently page table walks will no longer see the invalidated entry. However, at this point, the translation information has not been removed from any TLBs which may have been caching it! If the operating system, thinking that the physical address range is available, then maps some unrelated virtual page into the same physical address range, it will be possible for threads running on cores with stale TLB entries to illegally access that newly mapped memory! Any reads by threads expecting to access the original page will see unrelated data from the new page instead, and any writes aimed at the original page never reach their originally intended destination page; instead, they will clobber the data in the new page.

Failure to flush stale TLB entries is clearly a problem for correctness, but it is also a major problem for security as well. It breaks one of the key requirements of the VM abstraction: that each process see its own isolated virtual address space, and nothing else. If a thread were somehow able to delay a TLB invalidation, then it could orchestrate a side-channel attack in which it tried to take advantage of this behavior. For example, if a thread could deallocate a VM range without flushing its TLB, then it could simply continue to read from the now-unmapped virtual address range while waiting for some other process' page to be loaded into that page's original physical address mapping. It could then search for any sensitive data in that page: passwords, encryption keys, or anything else that might be considered private. The operating system therefore plays a critical role in ensuring that not even malicious processes are able to pierce through the VM abstraction.

The process by which stale TLB entries are invalidated is known as a *TLB shootdown*. The details of how TLB shootdowns are performed varies widely by architecture. At the most basic level, shootdown may be triggered via an explicit opcode in the instruction set, or it may be triggered as a side effect of accessing some control/status register (CSR). Many ISAs in fact provide multiple options which take effect at different granularities, as we discuss below. On some architectures, TLB shootdowns are even more involved. Each core may be able to invalidate any stale entries from its own TLBs, but for many of the same reasons that TLBs are not kept coherent, there may be no mechanisms for cores to directly invalidate the TLBs of *other* cores. Instead, on such architectures, cores which modify the page table are responsible for sending any remote cores a signal that they should do so themselves. We attack these issues one by one below.

6.2.1 INVALIDATION GRANULARITY

The first question is the granularity at which a TLB invalidation should be performed. If one particular page table entry is modified, then only that one translation needs to be invalidated from other cores' TLBs. To invalidate the entire TLB of each core would be overkill, as no other entries are made stale by the modification of just a single entry. Therefore, architectures frequently provide a mechanism by which TLB entries can be invalidated at the granularity of even just a single entry. On the other hand, a context switch on a TLB with no ASIDs (or under operating systems which do not use them) does generally require all of the non-global TLB entries to be flushed, and looping over an instruction which invalidates the TLB one entry at a time would add a lot of latency to the critical path of each context switch. Therefore, architectures also frequently provide an instruction to more efficiently invalidate the entire TLB (or all non-global TLB entries).

For example, on ARMv8 machines, the TLB is flushed using the `tlbi` instruction. This instruction can be qualified with the parameter `ALL` to flush the entire TLB, with the parameter `VA` to flush only entries matching the provided virtual address and ASID, or with a few other variants of the above as well. On older ARM processors, the TLB is flushed by writing a particular command value into the system control coprocessor (CP15) c8 register. On x86 machines, the TLB is flushed by writing to the `CR3` register, which stores the base pointer of the page table of the currently executing process. Linux implements this using the following code:

```
static inline void __native_flush_tlb(void)
{
    preempt_disable();
    native_write_cr3(native_read_cr3());
```

```
    preempt_enable();
}
```

Even though this writes the identical value back into the register, the TLB is still flushed as a side effect of the write. Note also that preemption is disabled during this time. This is because the read and write of the register are not atomic: if an interrupt were to arrive just in between the read and the write, and if the interrupt handler were to itself update CR3, then CR3 would be clobbered with the old value after returning to the original context. Once again, this shows how complex and subtle it can be to write thread-safe and reentrant synchronization code properly.

In between the two granularity extremes lies a gray area where the choice of granularity may be unclear. Suppose a medium-sized memory range is being deallocated from the virtual address space of a process. Is it better to iterate over the range one page at a time? Or is it better to just invalidate the entire TLB since that can generally be done with lower latency? In fact, this remains an open area of research and development, and often the answer is just to find some heuristic or threshold through profiling.

Consider the code snippet of Figure 6.1, which shows how Linux determines the page-by-page vs. global invalidation threshold. The threshold is simply set at a constant based on some basic intuition about how long the operation should take, how common large allocations are, and so on. This threshold will likely vary widely for each implementation, let alone for each architecture, especially with heterogeneity increasing with every generation (Chapter 7). It is therefore likely that heuristics such as this one will need to be revisited in the future.

Finally, architectures may provide any number of other hooks and/or variants of the above options in order to speed up the TLB shootdown process even further. Some architectures might provide a way to invalidate only non-global TLB entries, for example, so that global entries need to be invalidated only when the kernel modifies its own page tables. Architectures with software-managed TLBs will provide instructions to fill specific TLB entries as well. Each architecture (or even each platform) will deal with these specificities in its own way.

6.2.2 INTER-PROCESSOR INTERRUPTS

On some architectures a core can only invalidate entries from its local TLB. If TLB invalidations are kept local, then no hardware area needs to be specially dedicated for cross-core invalidation requests. Instead, invalidation messages are sent through other channels. Following a recurring theme, the choice varies from architecture to architecture. ARMv8 uses instructions which take effect across all cores, while IBM Power provides both a `tlbiel` instruction for invalidating the TLB of the local core and a `tlbie` instruction to invalidate the TLBs of all cores. The x86 architecture, however, does neither. Instead, it uses *inter-processor interrupt* (IPI).

An IPI is a particular class of processor interrupt that is sent directly from one core to another, or from one core to any subset of the cores on the processor—possibly to even to all cores, including the issuing core itself. The use of IPIs, as opposed to normal shared memory synchronization, ensures that the request is processed in a timely manner. Synchronization

```
/*
 * See Documentation/x86/tlb.txt for details.  We choose 33
 * because it is large enough to cover the vast majority (at
 * least 95%) of allocations, and is small enough that we are
 * confident it will not cause too much overhead.  Each single
 * flush is about 100 ns, so this caps the maximum overhead at
 * _about_ 3,000 ns.
 *
 * This is in units of pages.
 */
static unsigned long tlb_single_page_flush_ceiling __read_mostly = 33;
void flush_tlb_mm_range(struct mm_struct *mm, unsigned long start,
unsigned long end, unsigned long vmflag)
{
  /* skipping some code... */

    if (base_pages_to_flush > tlb_single_page_flush_ceiling) {
        base_pages_to_flush = TLB_FLUSH_ALL;
        count_vm_tlb_event(NR_TLB_LOCAL_FLUSH_ALL);
        local_flush_tlb();
    } else {
        /* flush range by one by one 'invlpg' */
        for (addr = start; addr < end; addr += PAGE_SIZE) {
            count_vm_tlb_event(NR_TLB_LOCAL_FLUSH_ONE);
            __flush_tlb_single(addr);
        }
    }

  /* skipping some code... */
}
```

Figure 6.1: The heuristic in arch/x86/mm/tlb.c used to determine whether Linux invalidates the TLB entries for a memory region page by page or in one global operation.

through shared memory would require the receiver core(s) to be explicitly checking or polling for requests, while the IPI approach ensures that the message is received and processed even if the receiving cores are not currently checking for invalidation requests.

The exact mechanism by which IPIs are performed is very specific to each system. Most processors do so through some form of hardware programmable interrupt controller (PIC) such as Intel's Advanced Programmable Interrupt Controller (APIC). Upon receiving an inter-processor interrupt, the receiving core traps into the operating system which in turn decodes it as an IPI request to perform a TLB shootdown. As such, it then takes the action of simply invalidating the local TLB as requested.

Algorithms 6.1 and 6.2 summarize the basics of two different approaches to performing a TLB shootdown using IPIs. These approaches are simplified versions of the corresponding Linux behavior. When a thread needs to perform a TLB invalidation, it executes the `tlb_shootdown` procedure. First, it invalidates its own local TLB. It then locks a per-CPU lock in shared memory; the releasing of this lock by the receiver will be used to indicate to the initiator that the receiver has completed the response. Note that the receiver need not send the response back as an IPI, as the initiator will in this case already be waiting on the lock, and so shared memory communication suffices here.

Algorithm 6.1 Basic TLB Shootdown Flow Using Per-CPU IPIs

```
function TLB_SHOOTDOWN()   // initiator
    invalidate_local_tlb();   // arch-specific
    for all cpu do
        if cpu != self then
            lock(get_lock(cpu));
            send_ipi(cpu, tlb_inval_func);   // arch-specific
    for all cpu do
        if cpu != self then
            while !unlocked(get_lock(cpu)) do
                spinloop_pause_hint();   // arch-specific

function TLB_INVAL_FUNC()   // receiver
    invalidate_local_tlb();   // arch-specific
    unlock(get_lock(cpu));
```

Next, the initiator sends TLB shootdown IPIs to the other cores in the system. The code of Algorithm 6.1 shows the IPI being sent to every other core, but the performance optimizations described below may filter this even further. Alternatively, it may sometimes be faster to simply broadcast a single IPI to all cores, possibly even including the issuing core, as shown in the variant of Algorithm 6.2. Meanwhile, the core receiving the TLB shootdown IPI executes the `tlb_inval_func` procedure: it simply invalidates its own local TLB, and then it unlocks

---

Algorithm 6.2 Basic TLB Shootdown Flow Using a Broadcast IPI

---

function TLB_SHOOTDOWN()   // initiator
    for all cpu **do**
        lock(get_lock(cpu));
    broadcast_ipi(tlb_inval_func);   // arch-specific
    for all cpu **do**
        if cpu != self **then**
            while !unlocked(get_lock(cpu)) **do**
                spinloop_pause_hint();   // arch-specific

function TLB_INVAL_FUNC()   // receiver
    invalidate_local_tlb();   // arch-specific
    unlock(get_lock(cpu));

---

its associated lock to indicate back to the initiating core that it is safe to proceed. Lastly, the initiator waits for each of the per-CPU locks to be unlocked, and once this occurs, it continues its execution.

6.2.3   OPTIMIZING TLB SHOOTDOWNS

TLB shootdowns are expensive; various characterizations estimate them to comprise anywhere from 5% to 10% of the runtime of a typical process, and up to 25% of the runtime of a virtualized process (see Section 7.4). They are also intrusive, as they interrupt many cores (if not all cores) and hence add a lot of overhead across the board. The problem is even further amplified when external devices are also mapped into a common address space with a CPU process (see Section 7.1). Because of their cost, engineers and researchers have considered many different ways to optimize TLB shootdown mechanisms.

As one common example, TLB shootdowns do not need to be sent every time a page table entry's accessed or dirty bit is set to true. Even in architectures with hardware page table walkers, the bits are still queried by the operating system by performing loads from the page table in memory, not by directly reading those bits from the TLB. Therefore, all OS page table management decisions will be made using the correct information. The worst that could happen if the TLBs are not invalidated after setting an accessed or dirty bit is that a another core may also take an unnecessary TLB miss that triggers a page table walker to again set one of the status bits, even though the bit was already set. Even so, this cost may be cheaper than doing a TLB shootdown after the first access, and so the benefits may be worth the complexity.

On the other hand, the OS must be careful to manage TLB invalidations when *resetting* the same accessed and dirty bits. The accessed bit is used to track whether a page is hot or cold, and as such, it must be able to detect new accesses to a page. If the TLBs are not invalidated after clearing the accessed bit, then subsequent accesses to the page may not re-trigger the accessed bit to be set again, and the page may be incorrectly considered cold. In this case, the cost of invalidating the TLB entry is outweighed by the need to get accurate profiling information.

It is even more critical to invalidate the TLB after resetting the dirty bit, as failing to do so would affect functional correctness. For example, suppose a dirty page has been flushed to its backing store and hence has become clean. However, suppose the TLBs were not invalidated after performing this flush. Any subsequent write to the page in question would pass through a TLB in which the page would still be marked as dirty. Therefore, thinking that there is no need to redundantly mark the page as dirty again, the TLB would not update the page table to mark the page dirty. In this situation, the VM subsystem would simply lose track of the fact that the page had been dirtied since its last flush to its backing store. This could easily then lead to data loss or data corruption. Due to both scenarios above, TLB entries are in fact invalidated when one of the status bits is reset.

For similar reasons, a TLB shootdown is not needed after upgrading a page's permissions. If a core whose TLB has not been updated accesses the page in a way that was illegal under the old permissions but is legal under the new permissions, then the core will simply take a minor page fault, invalidate the old TLB entry, and proceed. The page table walker will then fetch the newly updated page table entry. Just as above, however, it is not safe to do the same after a permissions downgrade. Instead, after a downgrade, invalidations must be performed.

Another important optimization deals with filtering the set of cores which must be sent a TLB shootdown request in IPI-based shootdown procedures. For example, many x86 processors do not have address space identifiers in their TLBs, or the ASIDs are not used, and hence context switches require all non-global (i.e., non-kernel) TLB entries to be invalidated on every context switch (Section 4.5.3). However, on such systems the OS needs to send shootdowns only to cores also working in the same context (unless the entry being invalidated is global). Cores working in different context can be filtered out. This trick can deliver huge benefits, especially as core counts continue to increase. DiDi went even further, proposing the creation of a hardware TLB directory to track the presence of page table entries in each TLB, thereby providing an even more accurate and higher-performing (albeit more expensive) filtering mechanism [110].

6.2.4   OTHER DETAILS

Before proceeding, we note two aspects of TLB shootdowns that we have glossed over so far. The first is the possibility that there might be synonyms that might also need to be invalidated. In particular, if the request is coming not via virtual address but rather by physical address, as would happen if a page frame were being swapped out to disk, for example, then all virtual addresses pointing to that page would need to be swapped out. Shooting down the TLB entries only within the same process and/or with the first matching virtual address may not be sufficient. Therefore, a proper TLB shootdown implementation might also need to perform the reverse map lookup and iterate through all of the virtual addresses pointing to the physical page frame in question.

The second aspect that we have glossed over so far is the fact that "TLB shootdown" is a bit of a misnomer. Besides the TLB, many architectures also have specialize page table walk caches of some kind, and these are also often incoherent with main memory (and with the TLBs, for that matter). On many architectures, even when they exist, they are often invisible to software. On such systems, they are simply invalidated whenever the TLB is invalidated. This does not affect the shootdown mechanism itself, but it does require the implementation of TLB invalidation commands to ensure that the page table caches are also invalidated as well. Recent versions of IBM Power, on the other hand, do provide an explicit page walk cache invalidation option.

6.3   SELF-MODIFYING CODE

When instruction caches exist independently of data caches, they are generally read-only, with updates expected to be relatively infrequent and hence not on the critical path. Like any other hardware structure, instruction caches are optimized for the common case. Therefore, just as for TLBs, many architectures do not dedicate hardware to keep instruction caches coherent. Furthermore, even if the caches themselves are coherent, a large number of instructions may have already been fetched and processed by the pipeline before the instruction memory update took place, and these in-flight instructions may not tracked by the coherence protocol.

Although updates to instruction memory may be rare, they do occur in a number of important real-world scenarios. For example, dynamic linkers often make use of a lazily updated procedure linkage table (PLT) which connects code to functions in dynamically linked libraries. Each update to the PLT is a write to instruction memory. Just-in-time (JIT) compilers also produce large code dynamically and then store it to memory so it can be executed. More broadly, self-modifying code may be used in general to perform any number of runtime optimizations or to implement various low-level debugging or profiling mechanisms. Nevertheless, the memory system must be able to correctly account for such updates when they do occur. Specifically, since the instruction caches and/or pipeline may have become incoherent, they must be flushed by software to ensure that the newly updated data will be properly fetched and executed.

The use of self-modifying code often requires operating system assistance due to policies such as W^X. While the code is being produced, it must be writable. However, when it is later executed, it must in fact be executable. Most modern machines prevent both permissions from being granted at the same time as a security precaution. Therefore, sometime in between the production and the execution of the code, the permissions of the region of memory storing the code must be switched from read-write to read-execute.

Following the theme, the mechanism for handling self-modifying code varies from architecture to architecture. x86 processors are designed to handle most cases automatically in hardware, requiring only that existing code jump into the modified code (as opposed to, say, modifying the next instruction and then naturally proceeding to execute it). However, there are two cases where a stronger synchronizing instruction (such as `CPUID`) is needed. First, if the self-modifying code is taking place in a multithreaded context, the remote thread must be notified in a manner very similar to a TLB shootdown, and a synchronizing instruction must be used in place of or in addition to the jump into the modified code. Second, if the modification is done through a virtual address synonym, then again, a synchronizing instruction is needed.

ARM and Power processors do not handle most self-modifying code automatically. Instead, they require the user (or in most cases, the operating system) to insert explicit specialized fences into ensure that modifications are propagated. Power processors use the sequence `dcbst;sync;icbi;isync`, which writes any changes to main storage, waits for the writes to have propagated, invalidates the instruction cache, and waits for the invalidation to have completed, respectively. ARMv8 uses the similar sequence `DC CVAU; DSB ISH; IC IVAU; DSB ISH; ISB`, which is analogous to the Power sequence, but which requires an additional `DSB ISH` between the latter two operations. (We study the memory fences themselves in the next section.)

Naturally, the various mechanisms used to enforce instruction cache coherence in the presence of self-modifying code can slow down execution significantly. Therefore, self-modifying code is not common, and it is not meant to be used frequently, particularly by user code. It is only used when the modification granularity is large enough that the cost is amortized away, such as would be the case for a JIT compiler optimizing a hot code path once then executing it repeatedly. When it is used, the user code and the operating system must take great care to ensure that coherence is restored through the use of the often-obscure but nevertheless very important instruction sequences described above.

6.4   MEMORY CONSISTENCY MODELS

The final topic we discuss in this chapter is the importance of accounting for memory consistency models when interacting with the VM subsystem. A memory consistency model (or simply a memory model, for short) provides a set of rules that defines what values may be legally returned by loads to memory. Intuitively, the intention of nearly every memory model is that each load returns the value of the latest store to the same location, just as we saw earlier in our definition of coherence. Unfortunately, defining what "latest" means, and coming up with a precise and complete memory model definition in general, has turned out to be a notoriously difficult problem. This is even more true when also attempting to reason about page table walks and instruction fetches, as these are even less coherent and even less strictly synchronized than even normal loads and stores.

Most modern processors implement some form of "weak" or "relaxed" memory consistency model in which memory accesses may be executed in an order that is different from how they were originally laid out in the code. For example, x86 architectures today use the TSO memory model shown in Figure 6.2. Most importantly, TSO allows a store to be reordered after a younger load. This in turn shortens the critical path of the load, which in turn usually shortens the critical path of the execution overall. In other words, stores are not generally on the critical path of the execution, and so they can be delayed more easily. Other architectures (such as IBM Power and ARM) go even further in permitting almost all memory accesses to be reordered by default. In each case, the architecture will provide one or more mechanisms (e.g., memory fences) to prevent the reordering of memory instructions that might otherwise get reordered. Such mechanisms are then used to build up synchronization primitives such as locks or mutexes that work properly even in the presence of reorderings that might otherwise take place.

The prototypical example of the need for ordering enforcement is shown in Figure 6.3. MP is the canonical example of a *litmus test*: a small program, boiled down to its most basic and abstract formulation, which is designed to test some property of a memory consistency model. Suppose [data] represents some data structure protected by the [flag] synchronization variable. The programmer's intention is that the producer will not write to [flag] until it has already written to [data]. Likewise, the consumer should not read from [data] until it has first

		Second	
		Load	Store
First	Load	X	X
	Store	—	X

Figure 6.2: A loose definition of the total store ordering (TSO) memory model used by x86. An "X" indicates that a pair of accesses of the corresponding types are never reordered; a "—" indicates that the corresponding pair may be reordered.

read the updated value of [flag]. Together, these constraints would ensure that the consumer can only read the updated value [data]=1 and not the old stale value [data]=0.

Producer Thread	Consumer Thread
`*data = 42;`	`while (*flag ! = 1) { /* loop */ }`
`*flag = 1;`	`int d = *data;`

Figure 6.3: Pseudocode for the "message passing" (MP) litmus test.

Now, suppose a performance-seeking optimization decided to reorder either the stores or the loads in MP. If the stores were reordered, then the consumer would be able to read from [data] before the producer would have updated it, thereby possibly returning the stale value [data]=0 to the consumer. Likewise, if the load of [data] were to be reordered with the load of [flag]=1, then again, the stale value [data]=0 could be returned to the consumer. Therefore, architectures or compilers which reorder memory accesses by default much provide a way (e.g., a memory fence) to prevent these reorderings where synchronization variables are involved.

6.4.1   WHY MEMORY MODELS ARE HARD

The key challenge in defining weak memory models is that in all but the most trivial microarchitectures, each core has a different perspective on the order in which memory events take place! This means that there is no single universal notion of "latest" which can be used to determine which value each load should return. Although the programmer intuition might be that stores to memory take place immediately and globally, in reality, store buffers, caches, queueing, or even cache coherence protocols themselves might buffer stores in ways that make them visible to some cores earlier than others.

Store buffers are well-established means of decoupling CPUs from memory stalls and for removing stores from the critical path of CPU execution, and nearly all CPUs have store buffer or equivalent. Memory models are therefore generally expected to allow any behaviors introduced by store buffering. When a CPU issues a store into its local store buffer, a younger load from the same core is generally allowed to forward its value directly from the store buffer. This saves the load an L1 cache round trip of latency.

However, now consider the effect a store buffer has on the visibility of stores. In the store buffer forwarding example just described, the issuing core obviously observes the store to take place before the load. However, every other core in the system will see exactly the opposite! The load's return value will have already been determined before the store ever leaves the store buffer and proceeds to memory, and therefore every other core in the system sees the load happening before the store. It is exactly this kind of situation that introduces complexity into memory consistency models. Any definition of "latest" must either specify the core whose perspective is used, or it must carve out special exceptions for scenarios such as store buffer forwarding.

In fact, many memory models introduce behavior which is far more complicated that what we have just described. Memory consistency models are a separate field of study in their own right; we do not attempt to dig any deeper into such memory models in this book. Other existing books provide a more complete discussion already [101]. Instead, we simply focus on how even relatively simple memory model features such as store buffering can introduce complexity into the VM subsystem.

6.4.2   MEMORY MODELS AND THE VIRTUAL MEMORY SUBSYSTEM

The first question that comes up when trying to merge VM with memory consistency models is whether the memory model rules apply to VM, to physical memory, to both, or to neither. These issues were first studied in detail by Romanescu et al., who proposed the use of different memory models for virtual and physical memory [94]. They concluded that virtual address memory consistency and physical address memory consistency were fundamentally different, as virtual address memory consistency must deal with problems such as synonyms, mapping changes, and status changes that are not applicable to physical memory alone.

The second major question deals with understanding how "special" loads such as page table walks and instruction fetches differ in behavior from "normal" loads. We have already seen how TLBs, page table caches, and instruction caches may be incoherent even when the rest of memory is kept coherent. This in and of itself introduces some new behavior. On top of this, page table walk memory accesses may themselves be reordered with respect to normal memory accesses, even in situations where such reorderings would otherwise be forbidden by the memory consistency model.

For example, on many architectures, a memory fence that enforces ordering with respect to a "normal" load may not enforce any ordering if the same load were replaced with a page table walk. Enforcing ordering with respect to page table walks often instead requires an even heavier-weight (i.e., higher-latency) fence that, e.g., invalidates the instruction cache, invalidates the TLBs, interacts with the page table walk FSM, or whatever else may need to be done to ensure the broader ordering enforcement that particular microarchitecture. Normal fences can often save some latency by not performing all of these operations, but such fences then cannot be used for synchronization of the VM subsystem.

As an example, the ARM architecture distinguishes between DMB and DSB fences. The former (in its most basic form) orders all preceding loads and stores with all subsequent loads and stores, but it does not order instruction fetches or page table walks. The latter does everything the former does, but it does also enforce ordering with respect to instruction fetches and page table walks. The DSB fence is slower than DMB, but nevertheless sometimes necessary. This now explains the DC CVAU; DSB ISH; IC IVAU; DSB ISH; ISB sequence we saw earlier for self-modifying code, as well as the sequence DSB ISHST; TLBI VAE1IS; DSB ISH; ISB needed to perform a shootdown after modifying the page table.

Unsurprisingly, modern VM subsystems are highly prone to some very subtle types of memory ordering bugs. Consider the example of Figure 6.4, which depicts an abstraction of a bug found in the Linux kernel in 2016 [78]. Linux uses a bitmask, represented by cpu_mask in the figure, to track the CPUs that are active in each process context. This bitmask is used to filter the set of CPUs that need to be send TLB shootdowns, as described in Section 6.2.3. That works in isolation, but it was prone to the following race condition.

Thread 0: Updating Page Table	Thread 1: Switching Contexts
`// Update the page table` (a) `*pte = new_pte;`	`// enter the new context` (c) `*cpu_mask |= (1 « CPU)`

`// Shoot down remote TLBs`
(b) `int m = *cpu_mask;`
`    for (int i=0; i<NUM_CPUS; i++)`
`    {`
`      if (m & (1 « i))`
`        shootdown(pte, i);`
`    }`
	`// clear the old context`
	`flush_tlb(); // and fence`
	
	`// start executing in`
	`// the new context, and`
	`// do a page table walk:`
	(d) `(load *pte into TLB)`

Figure 6.4: An example of the complex interactions between the virtual memory subsystem and the memory consistency model.

Suppose Thread 0 is in the middle of updating the page table of its current context: it updates the relevant page table entry, and then it shoots down any CPUs whose corresponding bits in the mask are set. Meanwhile, Thread 1 is switching from some other context into the same context as CPU 0: it sets its corresponding bit in the bitmask, it flushes its TLB of the entries from the previous context, and then it begins executing. Assuming it touches the page referenced by the page table entry just updated, it will at that point naturally load the page table entry into its TLB. As we have discussed, from the point of view of the memory system, this page table walk is just a special sequence of loads; we model it here for simplicity as just a single load.

The race condition lies in the interaction between the four labeled memory accesses. As we saw earlier, x86-TSO allows stores to be reordered after later loads as a performance optimization. In this case, `cpu_mask` is acting as a special type of synchronization variable, but *there is no fence between (a) and (b)!* Therefore, (b) could be reordered before (a), and in particular, this makes the interleaving (b), (c), (d), (a) legal. Under this interleaving, (d) takes place before (a), and hence Thread 1 gets the old value of the page table entry. However, Thread 1's CPU does not receive a shootdown, because (b) happens before (c), and therefore Thread 0 does not see the bit corresponding to Thread 1's CPU as being set. Therein lies the problem: the stale translation remains un-flushed in Thread 1's TLB. Unfortunately, subtle bugs of this flavor continue to appear somewhat regularly [77] and often lead to very serious security violations such as root-level privilege escalation.

In response, recent research has started to chip away at the conceptual gap between memory consistency models that address only "normal" loads and stores and other aspects of the VM subsystem. Pelley et al. introduced the notion of a memory persistency model, for memory that lives in persistent non-volatile storage as opposed to traditional volatile memory [87]. Likewise, Bornholt et al. introduced the notion of a consistency model for filesystems, which live on disk as opposed to memory [24]. Lustig et al. introduced the term transistency model for memory models that take page table entry status bits into account in ways that consistency models alone cannot describe [80]. The latter two introduce the Ferrite and COATCheck tools, respectively, each of which applies formal memory model analysis techniques to their respective domains.

6.5   SUMMARY

In this chapter, we discussed the types of complexity that can arise due to aggressive caching and buffering of memory accesses in the microarchitecture. Although store buffers, caches, and TLBs undeniably provide enormous performance benefits, the latency, power, and/or area cost of keeping all such structures always fully up to date is generally prohibitive. At best, the corner cases of these behaviors (e.g., page table updates, self-modifying code) are generally off of the critical path anyway, and hence they are left to software. This in turn can present a tremendously complicated and highly architecture-specific or even implementation-specific set of requirements that programmers must follow to ensure correctness. The complete details are beyond the scope of this book, but our intention is to provide an overview of the topic as a starting point for further exploration.

With that said, we now add yet another layer of complexity by studying how the VM subsystem works in the presence of architectural heterogeneity and virtualization.

CHAPTER 7

Heterogeneity and Virtualization

VM remains an active area of research and development. Although we have largely focused on more traditional single-processor, homogeneous-memory scenarios so far in this book, today's systems are becoming increasingly heterogeneous and increasingly diverse. A computer today may be built from a half dozen or more different types of compute unit, many specialized for specific tasks such as video decoding. Backing these architecturally heterogeneous components is an equally varied set of caches, buffers, and networks which comprise the memory system. As these peripheral components continue to become more sophisticated and more general-purpose, they are moving away from strictly relying on bulk transfers of physical memory as the primary communication mechanism. Instead, they are moving toward becoming fully fledged participants in the VM subsystem, thereby allowing them to take advantage of all of the benefits of VM, just as CPUs have done for decades.

Of course, an ever-broadening VM design space brings with it major new challenges. The complexity of today's VM implementations is growing exponentially due to the sheer number of dimensions that must be covered, including diversity in instruction set architecture, memory technology, memory layout, management policies, virtualization, and so on. Within this space, some devices (e.g., CPUs, GPUs, and DSPs) may communicate with each other via shared VM, while others (e.g., network cards, disk controllers) may communicate only via physical memory and DMA. (All of these are described below.) There may even exist virtual address ranges which are not meant to be backed by any memory at all! Ensuring the correctness of such systems presents a huge burden to today's designers, but it also presents an exciting new area of research that architects and system designers are only now beginning to explore.

Some of the specific challenges introduced by heterogeneity and virtualization include the following.

1. Memory accesses can have much higher latencies, when the data being accesses currently lives on another device or even on another machine.

2. The device's virtual address space may the same as, a subset of, or unrelated to some host process' virtual address space.

3. Handling page faults is difficult, as often there is no operating system running on an accelerator.

4. Implementing cross-device cache coherence is difficult and expensive, but omitting hardware coherence places a much larger burden on the programmer.

5. Optimal management (allocation, migration, and eviction) of data into a physical memory becomes a highly non-trivial problem in the face of diverse memory performance characteristics.

6. In virtualized systems, memory references and VM management must either pass through the hypervisor, adding an extra level of indirection, or must rely on dedicated hardware support to avoid that extra latency

7. ...and so on.

In this chapter, we present various advanced uses of VM and give a brief overview of how they attack some of the problems listed above. Some of these systems are already widely deployed commercially, while others are relatively new or still even just research prototypes. We do not attempt to cover the full depth and breadth of each topic. Instead, we give a broad survey of the state of the art, and we refer the reader to the cited references for more detail on each individual topic.

7.1   ACCELERATORS AND SHARED VIRTUAL MEMORY

While Section 2.4 discussed intra-processor shared memory scenarios, in a heterogeneous system it is also common to have situations in which a VM space is shared across multiple devices as well. A wide range of on-chip components today are moving toward shared VM, as the flexibility of VM enables more interesting use cases than the alternative of coarse-grained bulk task offloading can always provide. The ability provided by VM to share data structures with (virtual address) pointers between devices and the resulting elimination of any manual memory management burden are huge boons to the programmer. As a result, devices as broad as graphics processing units (GPUs), digital signal processors (DSPs), and the Intel Xeon Phi manycore chip now share a virtual address space with their host CPUs.

For example, consider the Qualcomm Snapdragon 800 series of SoCs. Each such chip contains three types of compute unit that are "[a]rchitected to look like a multi-core with communication through shared memory": the ARM CPU, the Adreno GPU, and the Hexagon digital signal processor (DSP) [93]. SoCs today regularly contain a dozen or more types of accelerators, and it is not unreasonable to believe that more and more of them will interact through shared memory as time progresses. The trend in GPUs' use of VM over the past decade is particularly enlightening, and so we use GPUs as our primary case study in the rest of this section. However, most of the discussion that follows applies equally well to other shared memory devices.

GPUs were originally specialized hardware meant solely for rendering graphics. Over time, as people began to realize their more broader potential, GPUs became more and more programmable and general-purpose, and users could program them almost as if they were just extremely parallel CPUs (with important caveats which are outside the scope of this textbook). As such, while early GPU computation models followed a bulk offload paradigm in which coarse-grained tasks are sent from the CPU master to the GPU peripheral, current GPUs share virtual address spaces, can to some extent manage their own memory and dynamically launch new tasks, and can even in some special cases share data with the CPU and with other GPUs at fine granularity. Likewise, the programming model has shifted from one in which all memory allocation had to be done in bulk using special allocation and deallocation commands, to one in which all mapped memory is accessible and managed automatically. Current research continues to further extend the frontier of the set of VM features that GPUs are capable of supporting.

One of the key limiting features today is the lack of system-level support for VM management, as compared to what CPUs can provide. For example, without an operating system on the GPU, there is no easy way to handle page faults. Either the system will simply trap, or the behavior will be simply undefined, or the request may send an (expensive!) message back to the CPU to handle the situation. Conversely, at least in the past there had been no easy way to maintain cache and TLB coherence when the caches, TLBs, and page tables lived on different devices—this was one major reason why today's implementations generally prefer to hand data off between devices one at a time. However, hardware support for coherence has begun to emerge more recently, on IBM Power + NVIDIA Volta systems and via AMD's Infinity Fabric, for example.

Even CPU operating system support for sharing VM across devices is somewhat limited. A notable example is the heterogeneous memory management (HMM) patchset for Linux [48]. HMM promises to provide a simple helper layer for device drivers to be able to mirror a CPU process' address space on an external device, so that memory that users allocate from the CPU is directly and easily accessible to the device. It also provides the necessary hooks and helper functions to migrate pages between devices, intercept and propagate synchronization operations, and so on. As of the time of writing, the HMM patches are just getting upstreamed into Linux 4.14 after many years of development.

GPUs in particular also generally provide extremely weak memory models which make communication more difficult to reason about and which make synchronization primitives more challenging to implement. GPU microarchitectures are designed to maximize throughput above all else, which means that the microarchitecture aggressively buffers, coalesces, and reorders memory requests, and interrupting that flow to perform reliable shared memory communication becomes difficult in the first place. They are also not always kept hardware-coherent with the CPU, which means that many of today's GPUs either don't allow the CPU and GPU to concurrently access any single piece of data in the first place, or they use some mechanism like on-demand page-level migration, which is slow, or they simply don't cache data that lives remotely [1]. Recent research has led to the development of formal memory consistency models for the Heterogeneous System Architecture (HSA), for the OpenCL programming language, and for the NVIDIA PTX intermediate representation for GPUs [52, 70, 83]; both are based on C/C++ but with explicit scope annotations.

7.2   MEMORY HETEROGENEITY

So far in this book, we have more or less implicitly assumed that all physical memory was homogeneous: that every region of physical memory delivers roughly the same performance characteristics (i.e., latency and bandwidth) to each core. However, in real systems this assumption often does not hold true. A system in which all of memory appears the same to all cores is known as a uniform memory access (UMA) system, while a system in which some regions of memory behave differently than others is known as a non-uniform memory access (NUMA) system. In a NUMA system, any memory access from a core to a DRAM on the same socket (or in the same cluster) will have relatively low latency, while any memory access from a core to a DRAM on another socket (or cluster) will have relatively higher latency, as it must traverse some kind of interconnect in addition to its original access path and latency.

Both UMA and NUMA systems are common today. A system may be UMA when all of its memory is physically located in the same place, as shown in the top left of Figure 7.1. It may also be UMA even when the memory is distributed, if the network that connects all of the components together has a uniform latency regardless of the source and destination, as depicted in the bottom left of the figure. On the other hand, systems can become NUMA in a number of different ways. Most commonly, the available physical memory in a system may be physically homogeneous but laid out in such a way that some regions of physical memory are farther away from some cores than other regions in terms of latency and/or bandwidth. This scenario is depicted in the top right of Figure 7.1. Lastly, it is also possible that the memory itself is actually physically heterogeneous, as the variance in performance characteristics of different memory technologies can itself be a source of heterogeneity. This scenario, shown in the bottom right of the figure, is discussed in Section 7.2.2.

7.2.1   NON-UNIFORM MEMORY ACCESS (NUMA)

Sometimes, the VM may be shared across multiple distinct nodes in a distributed system. Early distributed NUMA implementations were not cache coherent, and so they instead depended on overly complex programming models, making them somewhat more difficult to use. The Stanford DASH project was the first to build a scalable distributed cache coherent NUMA (ccNUMA) system [76]. DASH also led to the development of foundational principles such as release consistency, now a widely popular memory consistency model [46].

More commonly today, NUMA systems arise when memory is distributed across multiple processors within a single system. For example, many server-class systems today having multiple processor sockets on a single motherboard, and they use an on-board bus or an interconnect such as HyperTransport (originated by AMD) or Intel QPI to communicate between sockets. Single-system NUMA machines are generally kept cache coherent through the use of sophisticated

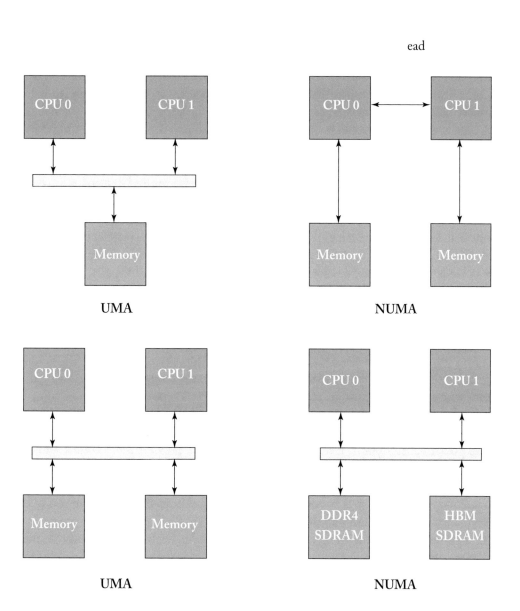

Figure 7.1: Examples of UMA and NUMA systems. In UMA settings, all memory is the same distance from the processor. In NUMA settings, the memory latency depends on the relative positions of the processor and the memory.

cache coherence protocols such as MOESI and MESIF which use Owned (O)/Forwarding (F) states to mitigate some of the need for heavy cross-socked communication. Lastly, single-system shared VM spaces are also being extended to peripherals such as GPUs.

The major new challenge posed by NUMA is the more complicated set of memory placement decisions that must be made to optimize the performance of the system. The simplest option is to simply ignore the fact that memory is heterogeneous, but this often leads to poor performance since memory may be placed unnecessarily far away from where it is being used. Instead, ideally all memory would be kept as close as possible to the core(s) accessing it in order to minimize latency and maximize throughput. For example, a system may choose to allocate memory regions as close as possible to the core performing the allocation, and then it may employ some kind of dynamic page migration policy to move pages around in response to changing system state.

Alternatively, for high-performance applications, the user may take more manual control of the application. Interfaces such as `numactl` on Linux allow users to inspect the hardware and software NUMA status, and then also to choose specific process affinities and memory allocation policies optimized for that application. The technologies, mechanisms, policies, and use cases for NUMA continue to evolve quickly and are likely to be important areas of research in the coming years.

7.2.2   EMERGING MEMORY TECHNOLOGIES

Orthogonal to the issue of memory placement is memory technology: a quickly growing number of systems today are making use of physically heterogeneous types of memory within a single system. High-end GPUs today are replacing traditional DDR SDRAM with Graphics DDR (GDDR) memory, which allows for better throughput characteristics at the expense of more aggressive power consumption. Even higher-end GPUs are moving past GDDR and exploring High-Bandwidth Memory (HBM), a 3D-stacked memory technology. These new types of memory can deliver very different latency, bandwidth, and storage capacity properties than traditional memory types, and so they add yet another dimension to the set of characteristics that a VM management policy must account for.

An alternative direction being explored is the use of persistent or non-volatile memory technologies—which hold their state even when the system is powered off—as memory as opposed to disk. Historically, non-volatile technologies such as NAND flash were too slow and had insufficient endurance to serve as memory as opposed to disk, but research prototypes and commercial products which narrow that gap are beginning to emerge. Intel and Micron's 3D-XPoint aims to sit right in between NAND Flash and DRAM, although it is not yet fully available in the market. Even more experimental technologies such as phase-change memory or even memristors are also being considered, but these are likely even further away [102].

Each new memory technology is likely to bring with it a new set of performance characteristics. Many non-volatile memories have much better read performance than write performance, both in terms of latency and in terms of endurance. Allocation and migration policies must account not only for the new parameters but also now for the asymmetry. Non-volatility in memory would be an even more fundamental change to the VM stack, as it would change the basic functionality of memory on top of just delivering different performance characteristics. Researchers have already begun to consider the implications of topics such as "persistency models," or consistency models for persistent memory [87], and some of this has begin to slowly make its way into industry as well. The latest ACPI specification, for example, defines memory zones in terms of latency, bandwidth, persistence, and cacheability [107].

7.3   CROSS-DEVICE COMMUNICATION

When the memory is distributed across more than one node, there needs to be some mechanism by which data is communicated between nodes when explicitly requested or implicitly needed. The most common mechanism by which that happens is DMA.

7.3.1   DIRECT MEMORY ACCESS (DMA)

So far, we have outlined the mechanisms by which CPUs and devices can share access to main memory. However, for the coarse-grained bulk offload scenarios in which many peripheral devices and accelerators are used, it would be extremely inefficient and wasteful to use up CPU cycles to actually perform memory accesses one at a time when copying larger chunks of data from one device to another. Instead, a mechanism known as direct memory access (DMA) allows peripheral devices to have direct access to memory without the need to directly involve the CPU. With DMA, the CPU or device programs a small state machine called a DMA engine with the parameters of the memory to be copied. The CPU or device is then free to perform other unrelated work while the DMA engine performs the copy. At the end, once the copy operation is complete, the DMA engine sends an interrupt to the CPU or device to indicate that the copy has been completed.

Sometimes, particularly in older generations, a device can't address the full range of memory that the CPU has access to. In such cases, to send data, the CPU would set up a buffer (sometimes called a "bounce buffer," or more generally just as an instance of double buffering) somewhere within the device's addressable region, then copy the data from its original location into the buffer, and then perform the transfer from that buffer. Conversely, to receive data, the OS would first receive data from the device in some buffer allocated within the device's addressable region, and then it would copy the data from that buffer into its destination. Often there may even be yet another copy to push the data from kernel space into user space or vice versa. High-performance networking implementations often go to great lengths to minimize such copies in order to avoid the latency that builds up with each step. Mechanisms such as Remote DMA (RDMA) attempt to push this even further by skipping all OS interaction entirely. Instead, RDMA simply copies the data directly into the virtual address space of a user program, and the user protocol relies on some software-level protocol to identify when messages have been successfully transmitted or received.

Another important aspect of DMA is that hardware does not always automatically guarantee to keep DMA operations coherent with CPU caches. In coherent DMA systems, a DMA request will automatically probe the cache coherence protocol, and if necessary, it will fetch or flush dirty data from the caches into memory before performing the DMA operation. For systems in which DMA is not coherent, these operations must be performed by software, and hence the operating system is responsible for ensuring that no conflicts occur by manually triggering writeback of dirty lines and preventing other code from interfering with the memory region during the DMA operation. Conversely, some high-performance implementations of DMA will inject data received from external devices (such as network cards) directly into the caches of cores waiting for that data in an attempt to minimize the latency that the core will see when trying to read in that newly received data.

One interesting case study for DMA: the Sony, Toshiba, and IBM collaborated on the Cell architecture, used notably in the Sony Playstation 3 [92]. Cell arranged various compute units in a ring-like arrangement: one master "power processing element" (PPE) and multiple "synergistic processing elements" (SPEs). Unlike most multicore systems, these processing elements did not share address spaces with each other. Instead, they were designed to work independently, and the programming model was to build a regular communication pipeline out of inter-element DMA transfers. Although Cell was capable of delivering very high bandwidth in this way, it was considered a very challenging chip to program, and its DMA-centric programming model has not caught on widely.

7.3.2   INPUT/OUTPUT MMUS (IOMMUS)

VM on accelerators is supported through the use of a special type of MMU known most generally as I/O Memory Management Unit (IOMMU). An IOMMU serves roughly the same purpose as a regular MMU: it translates virtual addresses (or "device addresses," if the device is operating within its own private virtual address space) into physical memory addresses. Just as for the CPU cores and the MMU, all accesses from the peripheral device to CPU memory that miss in the device TLBs (if present) pass through the IOMMU to get translated, or when things go wrong, to take a fault. IOMMUs generally have access to a page table managed by the host CPU and fill in their device translation caches just as TLBs are filled. The IOMMU is distinguished from a regular MMU by its location relative to the components it manages: while MMUs are tightly coupled to the CPU cores, the IOMMU, by virtue of also sitting on the CPU die, may be naturally somewhat distant from the accelerators. It also serves as a single central hub for all device translation requests, standing in contrast to the distributed set of MMUs associated with the cores of a multicore processor. This is depicted in Figure 7.2.

In general, IOMMUs enable peripheral devices to benefit from VM in much the same way that CPUs do through regular MMUs. Some features of IOMMUs, however, are more

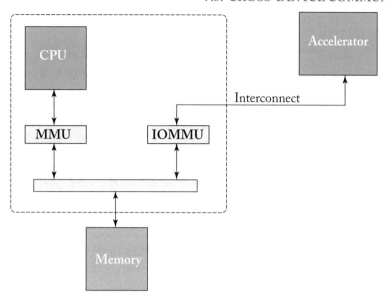

Figure 7.2: The IOMMU performs virtual memory management for peripheral devices and accelerators.

specialized. Many peripherals, particularly earlier or simpler devices, have fewer memory address bus bits than the rest of the system. In the past, devices would have to play tricks such as setting up bounce buffers, as described in the previous section. IOMMUs can also fill this gap by translating narrow device addresses into wider physical addresses in order to provide a more seamless method of integration.

One particularly important job for MMUs is enforcing isolation. If devices were given unrestricted access to physical memory, then it would be impossible to maintain the security of programs running on the machine. A malicious device could simply snoop on the memory of any running program in real time! The IOMMU is therefore responsible not only for providing VM functionality to external devices but also for ensuring that they do not access any regions of memory which have not been specifically made available to them by the CPU operating system.

Case Study: GPUs and IOMMUs

As a relevant case study, we focus on AMD x86-64 systems with GPU components, in a shared VM environment [84, 109]. Modern GPUs can maintain private TLB hierarchies for recently used translations. These are called device TLBs. However, unlike the CPU's MMU, the AMD IOMMU is not tightly integrated with CPU's data cache hierarchy (although this is an implementation choice and may change in future systems). The data caches may contain the most up-to-date translations but those cached copies cannot be directly accessed by accelerators. It is instead up to system software to ensure that any changes to the page table are propagated through the memory hierarchy until the point at which they become visible to the IOMMU and hence to the GPU.

On an GPU device TLB miss, a translation request in the form of a PCI-Express Address Translation Service (ATS) packet is sent over a PCIe-based internal interconnect to the IOMMU, which implements its own TLB hierarchy. If the translation request also misses in the IOMMU TLBs, a hardware page table walker in the IOMMU walks the page table. ATS requests are tagged with a process address space identifier (PASID) and the IOMMU maintains a table that matches PASIDs to page table base physical addresses. Once the address is successfully translated, the IOMMU sends an ATS response to the GPU. The protocol and packet formats for ATS requests and responses are part of the PCIe standard specification and are the same across all accelerators in the system.

Naturally, IOMMU page table walkers may, just like CPU IOMMUs, detect page faults. When this happens, the IOMMU sends an ATS response to the GPU notifying it of this failure. In response, the GPU sends another request called a Peripheral Page Request (PPR) to the IOMMU. The IOMMU places this request in a memory-mapped queue and raises an interrupt on the CPU. This foreshadows the need for OS intervention on the CPU to handle the page fault.

To improve performance, multiple PPR requests can be queued before the CPU is interrupted, enabling batched page faults. The OS uses its IOMMU driver to process this interrupt and the queued PPR requests. In Linux, while in an interrupt context, the driver pulls PPR requests from the queue and places them in a work-queue for later processing. This design decision was made to minimize the time spent executing in an interrupt context, where lower priority interrupts would be disabled. At a later time, an OS worker-thread calls back into the driver to process page fault requests in the work-queue. Once the requests are serviced, the driver notifies the IOMMU. In turn, the IOMMU notifies the GPU. The GPU then sends another ATS request to retry the translation for the original faulting address.

7.3.3   MEMORY-MAPPED INPUT/OUTPUT (MMIO)

Another special-case yet very important use of VM is memory-mapped I/O, or MMIO. In MMIO, some external communication mechanism, most commonly a configuration register for some external device, is mapped into the physical address space of some CPU process in place of some piece of actual physical memory. By mapping this "physical memory" into the virtual address space of a process, the process is able to communicate with the device or port using normal read and write commands, just as if it were reading and writing a normal memory location. In other words, MMIO eliminates the need for separate dedicated logic for I/O requests. One downside to MMIO is the loss of actual physical memory from the address space, but this is much less of a concern in modern systems than it used to be in the more constrained systems of the past.

For example, suppose some PCIe expansion card wishes to make its configuration accessible to the operating system driver on the CPU. One common way to do that is for the card to expose some set of registers and/or buffers as part of its PCIe configuration. The bootloader and operating system will then enumerate these buffers and registers during boot, and the driver will access this configuration as it is loaded to map the registers and buffers into the kernel's virtual address space. From that point on, the driver is able to communicate with the device simply by reading and writing the mapped addresses, according to whatever (often proprietary) protocol is used by the hardware.

Although the MMIO abstraction is convenient for the simplicity of the programming model and of the implementation, treating I/O requests as memory requests as memory request can present some gotchas. First of all, the requests must take special care to ensure that they are not stalled indefinitely in the cache rather than progressing straight to the MMU and then to the device; the ability to mark memory accesses as non-cacheable is discussed in the next section. But caches aside, it is important to keep in mind that while MMIO requests use the "language" of memory reads and writes, they can violate every other rule that memory normally follows. For example, MMIO reads may not return the value of the latest write to the same address, because in reality, the underlying "memory" is actually a device register or some other communication port. In fact, even reads are not harmless—a read of a device register can be destructive or trigger some sequence of side effects just as easily as an MMIO write can.

7.3.4   NON-CACHEABLE/COALESCING ACCESSES

As noted in the sections above, some special VM features such as MMIO must be careful to ensure that in spite of their use of memory-like read and write requests, their operations are not actually cached, buffered, or coalesced as freely as normal memory operations. As discussed in detail in Section 6.4.2, memory systems very frequently reorder, buffer, and coalesce memory operations into whatever dynamic order will minimize throughput and/or latency, to whatever extent is permitted by the ISA specification for that system. However, this behavior is incompatible with MMIO in particular. Sequences of duplicate operations may have particular non-redundant meanings when made to device registers, and the order is generally important. Furthermore, it is crucial that MMIO accesses bypass the caches so that they do not get stalled indefinitely while waiting to be naturally evicted by the cache.

In order to allow MMIO to work properly, most processors provide some mechanism for indicating that certain memory regions and/or certain memory accesses should be treated as uncacheable. x86-64 processors, for example, provide the older memory type range register (MTRR) mechanism as well as the newer page attribute table (PAT). Uncacheable memory accesses not only bypass the caches entirely; they also are generally prevented from taking any path or entering any buffer where they could possibly be reordered with other memory accesses. In fact, some MMIO accesses are treated as strongly as full memory fences in an effort to ensure compatibility with devices that impose strong ordering restrictions. As such, uncacheable accesses are generally extremely slow and disruptive to the memory system, and ideally their use is either minimized to the extent possible or at least mitigated through the use of techniques such as batching.

On the other hand, other types of special memory regions may have the opposite desire: they may wish to maximize throughput at all costs, even if it means temporarily giving up some normal memory functionality in the process. For example, consider the example of GPU framebuffers, which store information about the next frame that will be rendered onto the screen. Traditionally, framebuffers presented a very one-way communication paradigm: the CPU filled in polygons and pixels over time, and then when the process was complete, the frame was passed to the GPU to be rendered. Following this pattern, framebuffers follow a very one-way, write-only communication pattern. Because it knows reads are not important in that context, the memory system may have significantly greater flexibility to batch, reorder, and stall writes, bursting them out only at the end or when completely filled, in order to fully maximize the available throughput.

For scenarios like those presented by the framebuffer, many CPUs make special write-combining, write-coalescing, or non-temporal modes available for use in special situations. These regions typically work well for read-only or write-only scenarios such as those described, but any mixing of reads and writes will either simply return some unpredictable values or will impose a large penalty due to the need to fence instructions back into some more sensible order. For example, a read may not be able to quickly return the value of the latest store written to the same address. Instead, a non-temporal load may simply return whichever value simply happens to be the one most readily available, newest or not. Or, a load to a write-combining region may force the write-combining buffers to drain in order to ensure that the newest value is ultimately returned, even if it takes a long time to find it.

While special cacheability regions are not generally used for general-purpose user code due to their challenging programming model and often limited capacities, they play very special and important roles in keeping the system operating as efficiently as possible. They may bend the VM abstraction a bit, but the resulting throughput benefits and/or programmability improvements make them clear wins in terms of overall system performance and usability.

7.4   VIRTUALIZATION

Virtualization is the critical technology enabling cloud infrastructures today. Among its benefits are improved security, isolation, server consolidation, and fault tolerance. Unfortunately, its also presents performance challenges. Ideally, systems would run applications on virtual machines with the same performance as running the application natively. However, VM is one the primary contributors to a performance gap between native and virtualized application performance [43, 44, 91]. The main problem is that virtualization requires two levels of address translation. In the first level, a guest virtual address (gVA) is converted to a guest physical address (gPA) via a per-process guest OS page table (gPT). Then, the gPA is converted to a host physical address (hPA) using a per-VM host page table (hPT). There are two ways to manage these page tables: nested page tables and shadow page tables.

7.4.1 NESTED PAGE TABLES

Most virtualized systems use nested page tables. x86-64 systems use 4-level multi-level radix trees for both page tables [17, 44, 91]. We refer to these as levels 4 (the root level) to 1 (the leaf level) as per recent work [11, 17, 18]. When a process running in a guest VM makes a memory reference, its gVA must be translated to an hPA. Figure 7.3 shows this process. The guest CR3 register is combined with the requested guest virtual page or gVP (not shown in the picture) to deduce the guest physical page (gPP) of level 4 of the guest page table (shown as gPP Req.). However, to look up the guest page table (gL4-gL1), the gPP must be converted into the hPP where the page table actually resides. Therefore, we first use the gPP to look up the nested page

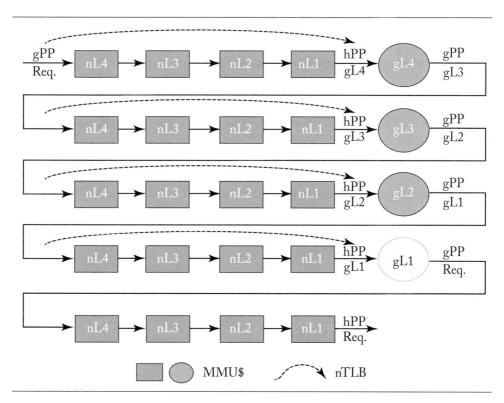

Figure 7.3: Two-dimensional page table walks for virtualized systems. Nested page tables are represented by boxes and guest page tables are represented by circles. Each page table's levels from 4 to 1 are shown. We show items cached by MMU caches and nTLBs. TLBs (not shown) cache translations from the requested guest virtual page (gVP) to the host physical page (hPP).

tables (nL4-nL1), to find hPP gL4. Looking up gL4 then yields the gPP of the next guest page table level (gL3). The rest of the page table walk proceeds similarly, requiring 24 memory references in total. This presents a performance problem as the number of references is significantly more than the 4 references needed for non-virtualized systems. Further, the references are entirely sequential. CPUs use three types of translation structures to accelerate this walk:

① Private per-CPU TLBs cache the requested gVP to hPP mappings, short-circuiting the entire walk. TLB misses trigger hardware page table walkers to look up the page table.

② Private per-CPU MMU caches store intermediate page table information to accelerate parts of the page table walk [10, 18]. See previous chapters for descriptions of MMU caching strategies.

③ Private per-CPU nTLBs short-circuit nested page table lookups by caching GPP to SPP translations [17]. Figure 7.3 shows the information cached by nTLBs. Concomitantly, CPUs cache page table information in private L1 (L2, etc.) caches and the shared last-level cache (LLC). The presence of separate private translation caches poses coherence problems. While standard cache coherence protocols ensure that page table entries in private L1 caches are coherent, there are no such guarantees for TLBs, MMU caches, and nTLBs. Instead, privileged software keeps translation structures coherent with data caches and one another.

7.4.2   SHADOW PAGE TABLES

Shadow paging is the alternative to nested page tables. With this approach, the hypervisor creates a shadow page table, which merges the gPT and hPT, holding a gVA to hPA translation.

TLB hits proceed similarly to nested page tables since the gVA can directly be translated to the hPA with no overheads. The main benefit of shadow paging, however, occurs on TLB misses. Unlike expensive two-dimensional nested page table walks, a shadow page table can be traversed with the standard 4 memory reference, since it maintains a direct path from the gVP to hPP.

The primary drawback of shadow paging is that page table updates are expensive [44]. The main problem is that shadow page tables must be kept consistent with guest and host page tables. In particular, guest page table updates can often be frequent, and suffer costly VM exits to update the hypervisor-managed shadow page table. Once the hypervisor is invoked, it invalidates or updates shadow page table entries. This mechanism can severely degrade performance since VM exits can cost 100s to 1000s of cycles.

7.5   SUMMARY

This chapter focused on the challenges facing VM in the context of hardware accelerators. We discussed TLB architectures suitable for GPUs and accelerators in general, whether the address translation hardware is placed within the device and/or in a centralized IOMMU. At a high-level, we made two important observations. First, VM support is becoming increasingly important for accelerators, in a bid to graduate from their otherwise pointer-restricted programming models and achieve programmability without excessively sacrificing performance. Second, the important challenge with many of these accelerators is that they have much higher address translation bandwidth requirements than traditional CPU designs. Consequently, components like page table walkers must be multi-threaded, while TLBs and MMU caches must be quite large.

This chapter also discussed the challenges imposed by virtualization. Virtualization is a particularly important concept for the design of VM since it is the bedrock on which cloud computing rests. Over the last few processor generations, vendors have added hardware support to accelerate memory translation and page table management operations for virtualization. We believe that a potentially interesting research question is how such support should be extended for the hardware accelerators that are now beginning to integrate support for address translation. We urge researchers to consider this important problem.

CHAPTER 8

Advanced VM Hardware

In the remainder of this monograph, we focus on advanced hardware and hardware-software co-design to reduce address translation overheads. This chapter discusses hardware techniques to achieve fast virtual-to-physical address translation. We cover hardware innovations for both native and virtualized systems.

There are several benefits to targeting hardware-based improvements. An important one is that hardware-only techniques that do not require changes to the software stack (i.e., the application, operating system, or compiler) generally enjoy faster adoption in commercial systems, since they require fewer changes through the VM layers. In other words, programmers can enjoy these performance benefits "for free."

The downside is that hardware approaches consume on-chip area and power, thermal, energy budgets. Generally, processor vendors are extremely judicious in adding/changing hardware, as on-chip resources could also be redirected to improving other portions of the microarchitecture. Additionally, all hardware must be verified for correctness. As we have already discussed in previous chapters, the VM hardware layer is particularly challenging to verify.

Therefore, while hardware changes can potentially offer high-impact solutions to the challenges facing VM today, they must also be sufficiently simple and readily-implementable. We now discuss a subset of the techniques that have been proposed by the computer systems research community in recent years. The discussion is non-exhaustive and is meant to serve as an introduction to recent research efforts.

8.1 IMPROVING TLB REACH

We begin by discussing hardware approaches to improve TLB reach. By TLB reach, we mean the effective capacity that a TLB offers. In other words, if we support a 1,024-entry TLB for 4 KB page translations, this corresponds to a total reach of 4 MB of memory. Naturally, the higher the reach, the lower the frequency of TLB misses.

8.1.1 SHARED LAST-LEVEL TLBS

Modern processors generally employ two-level TLB hierarchies per core. Therefore, L1 and L2 TLBs are usually private structures. However, recent studies [20, 79] have considered the potential benefits of designing last-level TLBs shared among multiple cores. While at first glance, there may seem to be parallels between shared last-level caches and a shared TLB, many key differences remain. First and foremost, shared TLBs cache at the page-level granularity rather than cache lines. This has interesting implications on the access patterns launched by different cores, and the likelihood that multiple cores access identical translations. Second, since TLBs are far smaller than caches, eviction and sharing play different and important roles in performance. Moreover, the penalty of a TLB miss is typically much more severe than a cache miss since an expensive page table walk is involved. Therefore, shared TLBs are not a simple extension of shared last-level caches.

Figure 8.1 presents a multicore chip with private, per-core L1 TLBs backed by a shared last-level L2 TLB. While this example uses just one level of private TLBs, more levels may be accommodated (for example, each core could maintain two levels of per-core private TLB followed by an L3 shared last-level TLB). Shared last-level TLBs are similar to last-level caches in that they are accessed on misses to the L1 structures. Figure 8.1 shows the shared last-level TLB residing in a central location, accessible by all the cores. While this centralized approach is a possible implementation (and is currently what has been studied in research papers), other distributed implementations are also possible.

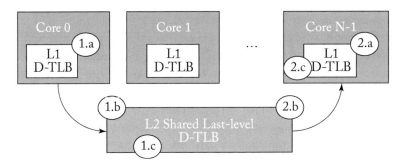

Figure 8.1: The basic structure of a shared last-level TLB involves a multicore chip with private, per-core L1 TLBs and a larger, shared L2 TLB [20]. Cases 1 and 2 detail instances of shared last-level TLB misses and hits, respectively.

Shared last-level TLBs enjoy two orthogonal benefits. First, they exploit the situation when multiple threads running on different cores access the same virtual-to-physical translations. As shown by past work [18, 20–22], this behavior occurs often for parallel programs where multiple threads collaborate to compute on shared data structures. In this situation, a core's TLB miss brings an entry into the shared last-level TLB so that subsequent L2 TLB misses on the same entry from other cores are eliminated. Second, even for unshared misses, shared last-level TLBs are more flexible than private per-core L2 TLBs regarding where entries can be placed. TLB hits arising from this flexibility aid both parallel and sequential workloads.

Figure 8.1 shows how shared last-level TLBs hits and misses operate. While these operations are numbered, there is no implied ordering between them. We detail the cases below.

Case 1: Figure 8.1 follows an L1 TLB and shared last-level TLB miss. In the first step, there is an L1 TLB miss (step 1a). Consequently, a message is sent to the shared last-level TLB. After the shared last-level TLB is accessed (which takes an amount of time equal to the access latency), we suffer a shared TLB miss (step 1b). The page table is then walked and the appropriate translation is inserted into both the shared last-level and L1 TLB. By entering the entry into the shared last-level TLB (step 1c), future misses on this entry are avoided by not just the core that initiated the page table walk, but also other cores that may require the same translation entry.

Case 2: Figure 8.1 also illustrates the steps in a shared last-level TLB hit. First, the L1 TLB sees a miss (step 2a), and a message is sent to the shared last-level TLB. Consider the case where there is a shared last-level TLB hit (step 2b). There are two possible reasons for this. The first is that the requestor core may have previously brought this entry into the shared last-level TLB. Alternately, the translation may be shared by other cores, so another core may previously have fetched it into the shared last-level TLB. Regardless of the reason, a shared last-level TLB hit avoids the page table walk. The same entry is now inserted into the L1 TLB (step 2c) so that future accesses to it can be quick L1 hits.

Having detailed its basic operation, we now discuss several important shared last-level TLB implementation options and details.

TLB entry design: Shared last-level TLB entries store information that is identical to the L1 TLB. That is, each entry stores a valid bit, translation information about virtual-to-physical mappings, and replacement policy bits. We also store the full context or process identifier tag with each entry. In other words, there is no real difference in the structure of entries between TLB types.

Multi-level inclusion: There exist many inclusion policies for private TLBs and the shared last-level TLB. Perhaps the simplest one is a mostly inclusive policy. In this approach (studied by prior work [20, 79]), page table walks result in fills into both the shared TLB and private L1 TLB. In other words, there is a best-effort approach to maintaining inclusion among TLBs, and enjoying the simplicity of management that this brings. However, mostly inclusive policies cannot guarantee inclusion, since the L1 TLBs and shared TLB maintain independent replacement policies. Therefore, it is possible for the shared TLB to evict an entry that remains resident in the L1 TLB, breaking strict inclusion. In this way, mostly inclusive TLBs mirror mostly inclusive caches. Just like caches, transitioning to strict-inclusion requires that TLBs also maintain back-invalidation messages. When the shared TLB evicts a translation entry, a back-invalidation message is relayed to the private TLBs so that the same translation can also be evicted from the private structures, if they exist there.

Finally, we can also make the TLB hierarchy exclusive. While inclusion is generally easier to design, it also wastes TLB capacity by duplicating translations in multiple TLB structures. Exclusive TLBs do not suffer from this waste. Shared TLB fills are performed only on L1 TLB evictions. In other words, the shared TLB becomes a victim cache of the L1 TLBs. This approach uses TLB capacity more judiciously than strict inclusion, but consequently has coherence overheads, as discussed next.

Translation coherence: Prior sections discussed the impact of TLB shootdowns and coherence. The organization of multi-level TLBs impacts the overheads of this coherence activity. One of the nice features of inclusive TLBs is that they simplify coherence management. Essentially, when a page table is modified, the shared last-level TLB can be probed to search for the corresponding translation. If the lookup misses, strict inclusion dictates that the private TLBs do not cache this translation either. Therefore, spurious lookups of the L1 TLBs can be avoided, a benefit that is not possible with mostly inclusive and exclusive TLB organizations. Overall, the utility of strict inclusion is dependent on the benefits of filtering coherence messages, vs. the overheads of back-invalidations. To date, such studies have not yet been performed, but remain an area rich for further exploration.

Centralized vs. distributed designs: The simplest shared TLBs are centralized and equidistant from all cores. This is feasible as long as the hit rate benefits outweigh access latency overheads. When shared TLBs become too big however, it may be necessary to use alternate designs to handle access latency. To understand this, consider shared TLB placement. As with caches, a communication medium exists between cores and the shared TLB (eg. an on-chip network or bus). Therefore, shared TLB roundtrip latency is comprised of the network traversal and shared TLB access time. Techniques that reduce on-chip network communication latency will therefore amplify shared TLB benefits. Moreover, it may be beneficial to consider shared TLBs that are distributed with non-uniform access latencies, similar to NUCA caches. Such studies remain open questions for the research community.

Integrating prefetching strategies: Past work has also considered augmenting vanilla shared TLBs with simple prefetching extensions [20]. Several studies have shown that due to large-scale spatial locality in memory access patterns, TLBs often exhibit predictable strides in accessing virtual pages. These strides occur in memory access streams from a single core [65, 97] as well as between multiple cores [22, 79]. While sophisticated prefetchers have been proposed to exploit this, only simple stride-based prefetching has been explored in the context of shared TLBs thus far. Specifically, on a TLB miss, past work [20] inserts the requested translation into the shared TLB and also prefetches entries for virtual pages consecutive to the current one.

An important design decision in prefetching is that it is critical to ensure that prefetches do not require extra page table walks. To avoid this, studies piggyback prefetching on existing TLB misses and their corresponding page table walk. When page table walks occur, the desired translation either already resides in the hardware cache or is brought into the cache from main memory. Because cache line sizes are larger than translation entries, a single line maintains multiple translation entries. For 64-byte cache lines, entries for translations for 8 adjacent virtual pages reside on the same line. Therefore, past work prefetches these entries into the shared TLB, with no additional page walk requirements. Moreover, all past work permits only non-faulting prefetches, so that there are no page faults initiated by prefetches. Naturally, all these design decisions remain ripe for further study and optimization.

8.1.2 PART-OF-MEMORY TLBS

Beyond shared TLBs, recent work has considered the design of extremely large shared TLB structures that are placed directly within main memory. These "part-of-memory" TLB designs [96] essentially leverage the following observation: while conventional wisdom dictates that TLBs must always be designed for quick access, in reality it is useful to design extremely large last-level TLBs with massive reach as long as the bulk of the TLB accesses are serviced from faster L1/L2 TLBs placed closer to the core. This approach marries both fast access time and large TLB reach and is particularly helpful in situations where TLB miss penalties are severe. For example, prior work [96] considers part-of-memory TLB designs in the context of virtualization with two-dimensional page table walks, where a TLB miss can take several tens to hundreds of cycles to service. These designs treat part-of-memory TLBs as hardware-managed versions of software Translation Storage Buffers (TSBs) discussed in previous chapters.

Figure 8.2 illustrates the general operation of a part-of-memory TLB. With this design, main memory is carved out into two separate regions. The first region remains the same as conventional memory, holding program instructions, data, and metadata in the form of page tables, and OS structures like the process control block, shared libraries, etc. The second region, however, is dedicated to the part-of-memory TLB. In other words, the physical address space is partitioned into a portion dedicated to the TLB and a portion for everything else. Carving out portion of main memory for a dedicated TLB also has the following advantage: it is possible to fill entries from the part-of-memory TLB into the hardware data caches, further accelerating their access.

To understand the mechanics of this organization, consider Figure 8.2. For the purposes of this example, we assume that each core maintains a single private TLB (although the design could accommodate private L2 TLBs without changing the logical flow of the lookup operation). In step 1, core 0 probes its TLB for a virtual-to-physical page translation. Ordinarily, a TLB miss at this step would invoke the hardware page table walker, which would walk the page table to identify the desired virtual-to-physical translation. However, a system that implements a part-of-memory TLB takes a different approach. Since translations (not to be confused with the page table itself) from the part-of-memory TLB are mapped in the physical address space, they can be resident in the hardware caches. Therefore, the first step is to look up the cache hierarchy for the desired virtual-to-physical translation. To construct the physical address to perform the lookup, we note that the part-of-memory TLB is resident in a specific portion of the physical address. This base address is added to the requested virtual address (see prior work for more details [96]), and the resulting sum is used to look up the cache hierarchy. Figure 8.2 assumes that we cannot find the desired translation in the L1 cache or LLC (steps 2 and 3). Consequently, in

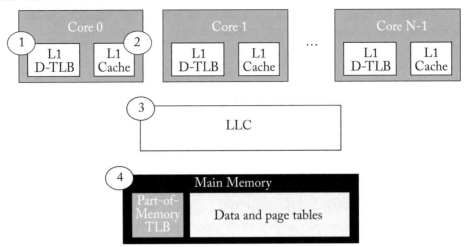

Figure 8.2: Main memory is carved into a dedicated portion for the part-of-memory TLB, and a separate portion for the remaining data, which consists also of the page tables [96]. Part-of-memory TLBs are accessed after private TLBs miss, but first involve lookups in the cache hierarchy.

step 4, we look up the part-of-memory TLB. At this point, a hit in the part-of-memory TLB results in a fill of the hardware caches and the TLB.

Alternately, it is also possible that the part-of-memory TLB does not have the desired translation. When this happens, a part-of-memory TLB miss message is relayed to the core. At this point, the core activates its hardware page table walker. From this point onwards, the standard page table walk process commences. For example, x86-64 systems with 4-level radix trees make four sequential memory references for the levels of the page table. These references may hit in the hardware caches or require main memory lookup, as is usual.

To better understand the structure of the part-of-memory TLB, consider Figure 8.3. Part-of-memory TLBs reside in DRAM banks. However, unlike page tables, which also reside in DRAM, part-of-memory TLB DRAM rows maintain the virtual to physical page mapping directly, rather than storing information from all four levels of the multi-level radix-tree page table. Further, they maintain a process identification number, permission bits, etc. Since DRAM is read in the unit of cache lines, several translation entries from the DRAM row holding the part-of-memory TLB are read out at once.

We point the interested reader to the original paper on part-of-memory TLBs [96] for more details on design issues such as supporting multiple page sizes, managing DRAM scheduling, and the memory controller, etc.

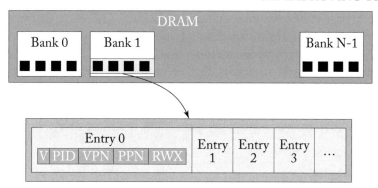

Figure 8.3: Structure of the part-of-memory TLB [96]. Each DRAM row maintains multiple TLB entries, which are carved out into a different portion of the physical address space.

8.1.3   TLB COALESCING

Shared TLBs and part-of-memory TLBs are approaches that present new design points in the organization of hierarchical TLBs. However, there are also other approaches that can change the design of any TLB (whether it is at the L1, L2, or shared level) to improve reach. One example of this class of orthogonal techniques is that of TLB coalescing [90].

The basic idea behind TLB coalescing is to augment traditional TLB hardware to introspect more intelligently on the activity of the application and OS to deduce patterns in page tables. The insight is that at the software level, the fluid interactions between many key stakeholders influences virtual-to-physical mappings. These include the activity of the application (i.e., how does it make memory allocation calls, how much memory does it request and in what order), other co-running applications (i.e., how much memory interference is there), and the OS (i.e., how effectively is it defragmenting memory). This means that page tables often (though they do not have to) see interesting patterns in virtual-to-physical translations. However, traditional TLB hardware is rigid, mapping a single virtual page to a single physical page, and is completely agnostic of these patterns.

TLB coalescing enables hardware that learns these patterns by using simple compression schemes to record information about multiple translations in a single hardware entry. An important design goal is to ensure that the hardware is kept simple so that there no need for additional area, and the reach increases. Further, with this approach, there is no need to modify software. TLB coalescing is not the only way in which hardware can introspect and learn page table patterns. For example, prior work has also considered TLB sub-blocking [103] while more recent work considers TLB clustering [88], which are similar in spirit. However, since TLB coalescing is today available on commercial systems like AMD's Ryzen chip, we discuss it in detail.

Figure 8.4 illustrates a high-level overview of TLB coalescing. On the left, we show a conventional TLB, caching information from a page table. Conventional TLB designs expect there to be no patterns in page tables, meaning that each individual TLB entry maintains a simple virtual to physical page mapping. Figure 8.4 shows, however, that in reality, page tables tend to see patterns. One common pattern is where multiple contiguous virtual pages are assigned contiguous physical pages. In response, a coalesced TLB can compress all these contiguous translations into a single TLB entry. As long as each coalesced TLB entry remains roughly the same size as a traditional TLB entry, the reach of the coalesced TLB quadruples in this example. We now shed more light on the sources of contiguity, and various hardware design options.

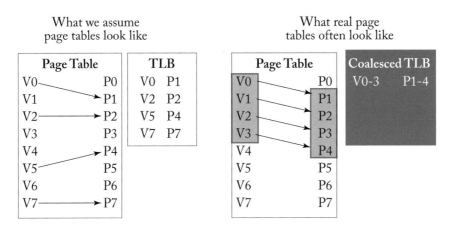

Figure 8.4: On the left, we show a traditional TLB caching four translations from a page table [90]. On the right, we show a page table with contiguous translations, and a coalesced TLB which uses one hardware entry to cache information about all four coalesced translations.

Sources of intermediate contiguity: In some sense, the notion of page table contiguity is something that we have visited in prior sections of this synthesis lecture, specifically when discussing superpages. To recap, superpages represent page sizes much larger in capacity than baseline page sizes. For example, x86-64 systems maintain 4 KB base page sizes, but also 2 MB and 1 GB superpages. Superpages represent situations with massive amounts of contiguity; i.e., 512 contiguous translations for 4 KB pages constitutes a 2 MB page (with the additional caveat that the pages be aligned at 2 MB address boundaries). 1 GB pages represent an even greater amount of this contiguity.

What distinguishes coalesced TLBs from superpages however, is that they focus on intermediate contiguity, or contiguity levels that have no alignment restrictions and are smaller than the 512 required for 2 MB superpages or 262,144 required for 1 GB pages. In practice, there are often situations where the software generates these levels of intermediate contiguity, which are not sufficient for superpage use, but remain useful as TLB coalescing candidates. Generally, there are three software agents that determine intermediate contiguity.

① The memory allocator: Consider the situation shown in the diagram on the left of Figure 8.5, where the application makes a memory allocation request for a data structure that occupies four pages. Naturally, this request is for four virtual pages, that will ultimately be used by the application. As shown in Figure 8.5, the memory allocator initially reserves four contiguous virtual pages (see our discussion on VMA regions in prior chapters), but does not yet assign physical page frames to them since none of the virtual pages have been accessed yet. Nevertheless, the first requirement of contiguity is met: contiguous virtual pages are reserved.

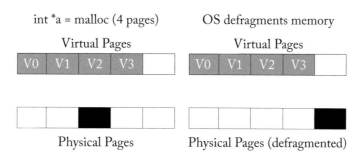

Figure 8.5: Memory allocation calls from the application and OS-level defragmentation contribute to intermediate contiguity formation.

② The memory defragmentation engine: Figure 8.6 also shows, in the diagram on the right, that the OS also periodically inspects the state of physical memory. At any point, physical memory may be heavily fragmented, as other processes enter and exit the system, and as system load generally increases. In Figure 8.5, we show that four physical page frames are free (in white) while one physical page frame is assigned (in black). However, the black frame fragments the free space by residing in the central frame number. Consequently, real-world OSes often run dedicated kernel threads to defragment memory. Figure 8.5 shows that the contents of the used frame are shifted to the end, so that all four free physical frames become contiguous. At this point, there is contiguity in the virtual address space, and the *potential* for contiguity in the physical address space.

③ The application: In the final step, the application makes memory references to the various virtual pages that it has been assigned. The first access to each individual virtual page generates a minor page fault (see previous chapters). If, as shown in Figure 8.6, the virtual pages are accessed in order (which many real-world applications tend to do), they are assigned physical pages contiguously.

The net effect of the interactions among the allocator, application, and OS defragmentation threads is that real-world applications often (though they do not have to) experience moderate amounts of contiguity. For example, past work has shown [90] that real-world workloads often see an average of 8–16 page table entries experiencing contiguity in the manner described.

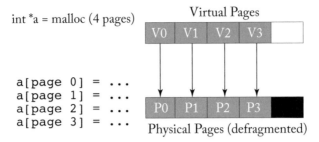

int *a = malloc (4 pages)

```
a[page 0] = ...
a[page 1] = ...
a[page 2] = ...
a[page 3] = ...
```

Figure 8.6: Application references lead to lazy page faulting patterns which contribute to intermediate contiguity formation.

This effectively means that by coalescing these entries in hardware, we can potentially boost TLB reach by 16× without additional storage area.

Coalesced TLB design options: Having discussed the sources of intermediate contiguity, we now focus on ways to exploit it using coalesced TLB hardware. We separate the cases of TLB lookup from TLB fills.

① Lookup: Figure 8.7 illustrates the process of a TLB lookup. On the left, we show a page table, where the virtual-to-physical page translations for V0-V3 are contiguous. On the right, we contrast the lookup for a traditional set-associative TLB with that of a coalesced TLB. We focus on a situation where a CPU probes the TLB for the translation for V1.

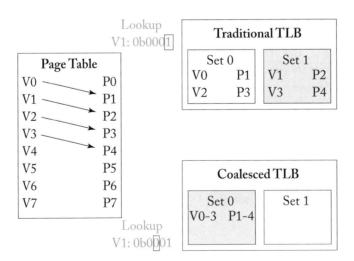

Figure 8.7: Lookup procedure for a coalesced TLB [90]. For a set-associative TLB, coalesced TLBs require a slightly modified set indexing scheme.

Conventionally TLBs use the lower-order bits of the virtual page number to identify the desired set number of the TLB to look up. In our example with a two-set TLB, this means that the lower-order bit becomes the index. For V1, this implies that the translation must reside in set 1. However, coalesced TLBs operate differently. Since we want to coalesce multiple contiguous translations into a single set, the set indexing scheme must permit translations for consecutive virtual pages to map to the same set. Suppose that we want to realize a coalesced TLB that supports a maximum of four coalesced translations. In this case, shown in Figure 8.7, we need translations for V0–V3 mapping to the same set, instead of striding across the two sets. Therefore, instead of using the lowest order bit as the index, this means that we must use bit number 2 to select the index.

Once the set index is known, we must scan the translations residing in the selected set. Figure 8.8 shows how this proceeds. Although we only show one resident translation, naturally, all translations in the set must be compared. Each translation maintains a bit vector that records which of the four possible coalescable translations are actually contiguous and hence coalesced. Therefore, a lookup requires both the conventional tag match using the virtual page bits, and also a lookup of the relevant bit in the bitmap. For a lookup of V1, this means that we must lookup the bit shown in blue. If this is set, it means that the requested translation exists in this bundle. The physical page frame can then be calculated by adding the relevant offset to the physical page frame number stored (P1 + 1 in our example).

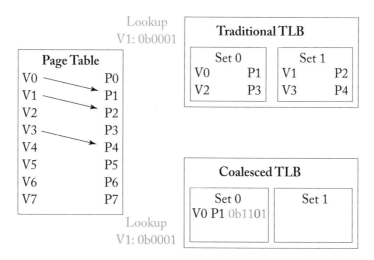

Figure 8.8: A bitmap encodes the existing translations in the coalesced bundle [90].

Figure 8.8 also implies several key aspects in the design of coalesced TLBs. First, it is possible for multiple translations with the same virtual page tag to co-exist in the set. For example, if V0–V1 were contiguous individually (by pointing to P1–P3) and V2–V3 were also contiguous (by pointing to, for example, P5–P6), set 0 would contain two separate entries for each of these two bundles of coalesced translations. The lookup distinguishes between these two coalesced bundles by looking up the relevant bit in the bitmap field. In the first entry, the bitmap would be 0b1100, while the second entry would store 0b0011. Therefore, the bitmap field essentially acts as a way to distinguish among multiple equivalent virtual page tags.

Second, the use of a bitmap also enables an optimization that goes beyond strict contiguity in the page table. For example, Figure 8.8 shows a situation where the translation for V2 does not map to P3. However, because V3 maps to P4, and this follows the contiguity pattern of the V0-V1 translations, it is possible to use a single entry to store all contiguous translations for V0, V1, and V3. The only caveat as the bitmap entry for V2 (shown in red) must be 0.

Finally, this design implies that the set indexing scheme has to be statically changed. As long as the page tables demonstrate sufficient contiguity, this is a good design choice. However, if page tables see little contiguity, such a change in the set indexing scheme can potentially increase TLB misses. Therefore, past work considers options such as using parts of the TLB real-estate for coalescing, while leaving the rest as traditional TLB hardware [88]. Other approaches consider coalescing only at the L2 TLB, leaving L1 TLBs unchanged [90]. Ultimately, the prevalence of real-world intermediate contiguity means that this remains a useful optimization, ushering its adoption in modern chips.

② Fill: We now describe the mechanism used to identify contiguous translations and fill them into the TLB. To ensure that TLB lookup latencies remains unchanged with coalescing, all coalescing is performed on the TLB fill path, which is off the critical path of lookup.

Figure 8.9 shows TLB fill and coalescing. On lookup, suppose that there is a TLB miss for the translation for V1. This triggers a TLB miss on both traditional and coalesced TLBs. In both cases, the page table walker requests portions of the page table in the unit of caches lines. Since multiple page table entries can reside in a cache line, this presents an opportunity for coalescing. Consider an x86-64 system, where 8 page table entries reside in a 64-byte cache line. With coalesced TLBs, combinational logic is added on the fill path to detect contiguous translations within the cache line. Once the contiguous translations are detected, they are coalesced and filled into the coalesced TLB. Past work [88, 90] performs several studies on the overheads of such combinational hardware and proposes ways in which they can be designed to be fast and simple.

In summary, the benefits of coalesced TLBs are twofold. The first is that they increase effective TLB capacity by caching information about multiple translations in a single entry. That this is achieved without any real area or access latency increase means that TLB performance is improved substantially. Second, and perhaps more subtly, TLB coalescing also performs a form of prefetching. For example, in Figure 8.9, the coalesced TLB bundles translations for V0, V2, and V3 and fills them into the TLB, even though the original request for V1 only. In the near future, accesses to V0, V2, and V3, which would otherwise have been misses on conventional TLBs, become hits. This form of prefetching also provides performance benefits.

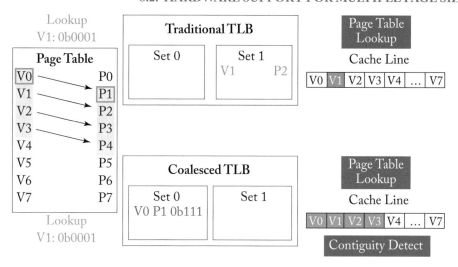

Figure 8.9: Coalescing is performed by combinational logic which inspects a cache line holding the page table entries and coalesces them on TLB fill [90].

8.2   HARDWARE SUPPORT FOR MULTIPLE PAGE SIZES

In previous sections, we discussed how commercial systems support multiple page sizes. Different approaches are taken for L1 TLBs, which are optimized for fast lookup, and L2 TLBs, which are optimized instead for capacity. As already discussed, most vendors use split or statically partitioned TLBs at the L1 level, one for each page size. This sidesteps the need to know the page size on lookup. A virtual address is used to look up all TLBs in parallel. Separate index bits are used for each TLB, based on the page size it supports; e.g., the set indices for split 16-set TLBs for 4 KB, 2 MB, and 1 GB pages (assuming an x86 architecture) are bits 15-12, 24-21, and 33-30 respectively. Two scenarios are possible. In the first, there is either hit in one of the split TLBs, implicitly indicating the translation's page size. In the second, all TLBs miss.

As we have already discussed, split L1 TLBs suffer from the key drawback of poor utilization. When the OS allocates mostly small pages, superpage TLBs remain wasted. On the other hand, when OSes allocate mostly superpages, performance can be (counterintuitively) worse because superpage TLBs thrash while small page TLBs lie unused. In response, there have been several proposals to support multiple page sizes concurrently. While some of them have partially been adopted by commercial products at the L2 TLB level, which can afford slightly higher lookup times, L1 TLBs remain split. We now discuss several research strategies to mitigate this problem.

8.2.1 MULTI-INDEXING APPROACHES

The first set of approaches, like split TLBs, essentially use different index functions for different page sizes.

Fully associative TLBs: The simplest strategy to accommodating multiple page sizes concurrently in a single TLB is to implement it with full associativity. In this approach, each TLB entry maintains a mask based on the page size of the translation. On lookup, the mask is applied to the virtual page (typically using a bitwise AND). The resulting bit string is then compared to the tag stored in the TLB entries. Naturally, the main problem with fully associative TLBs is that they have high access latencies and consume more power than set-associative designs. Therefore, the majority of emerging processors, particularly for servers, use set-associative TLBs. For example, Intel's Skylake chips use 12-way set associative L2 TLBs.

Hash-rehash: Since we desire set-associativity, we need to investigate alternate approaches to caching multiple page sizes concurrently. Hash-rehashed TLBs represent one such approach. With this approach, we initially look up the TLB using a "hash" operation, where we assume a particular page size. In vanilla hash-rehash implementations, this page size is typically the baseline page. On a miss, the TLB is again probed using "rehash" operations, using another page size. This process continues until all page sizes are looked up, or a TLB hit occurs, whichever happens first.

Hash-rehash approaches do enable concurrent page size support on set-associative TLBs but suffer from a few drawbacks. TLB hits have variable latency, and can be difficult to manage in the timing-critical L1 datapath of modern CPUs, while TLB misses take longer. One could parallelize the lookups but this adds lookup energy, and complicates port utilization. Consequently, hash-rehashing approaches are used in only a few architectures, and that too, to support only a few page sizes (e.g., Intel Broadwell and Haswell architectures support 4 KB and 2 MB pages with this approach but not 1 GB pages [26, 51]). That being said, recent work considers using prediction to ameliorate some of these issues [86]. We discuss such approaches later in this chapter.

Skewing: An alternative is to use skewing techniques instead. Skewed TLBs are inspired by skewed associative caches [32, 86, 99]. A virtual address is hashed using several concurrent hash functions. The functions are chosen so that if a group of translations conflict in one way, they conflict with a different group on other ways. Translations of different page sizes reside in different sets. For example, if our TLB supports 3 page sizes, each cacheable in 2 separate ways, we need a 6-way skew-associative TLB.

To elaborate, we now describe the principle operations of a skew-associative TLB. We refer readers to the original paper [99] for more details. The three fundamental design aspects of a skew-associative TLB are the following.

① For any given way I in the TLB, for any virtual page number, there exists a single entry E in way I and there exists at least one possible page size s, such that the virtual page number and s can be mapped on entry E in way I.

② Different hashing functions are used to index into the different ways of the cache.

③ If an application uses s as its only page size, it can still use all the TLB entries for translations.

Of these properties, ① is the definition of a skew-associative caching structure. The main problem with achieving it is that we do not know the virtual page associated with a virtual address at lookup, because it depends on page size. Consequently, skew-associative TLBs also enforce the following constraint.

④ Consider a function S, which we call a page size function. For any given way I in the TLB and for any given word at virtual address A, virtual page V(A) of word A can be mapped on way I if and only if the size of the virtual page is S (A, I).

In other words, if the virtual page V(A) is mapped on a way I from the TLB, then its page size can be automatically inferred. We refer readers to the original paper for detailed examples on skew-associative functions [99], and focus here on the higher level advantages and drawbacks of this approach.

On the one hand, skew-associativity does achieve the idea of caching multiple concurrent page sizes simultaneously in a TLB structure. On the other hand, its relatively complex implementation can cause problems. Lookups expend high energy as they require parallel reads equal to the sum of the associativities of all supported page sizes. One could hypothetically save energy by reducing the associativity of page sizes, but this comes with the cost of decreased performance. Finally, good TLB hit rates require effective replacement policies; unfortunately, skewing breaks traditional notions of set-associativity and requires relatively complicated replacement decisions [99]. Because of these problems, we know of no commercial skew-associative TLBs.

8.2.2 USING PREDICTION TO ENHANCE MULTIPLE INDICES

The common thread among the approaches for multiple page sizes discussed thus far is that we require multiple hash functions and lookups for the different page sizes. Naturally, multiple lookups increase the energy and access time overheads of this approach. Recent work has responded with clever techniques to reduce the need for these lookups by predicting, before looking up the TLB, what the page size of a virtual address is likely to be [86]. With this approach, the hash-rehash or skew TLB is first looked up with the predicted page size. Only on misses are the other page sizes then used. When prediction is accurate, this approach lowers the average TLB hit latency and lookup energy, by first looking up with the "correct" page size.

Consequently, prior work studies hardware that can be used to predict page size [86]. While the approaches they study are generalizable to predicting among several page sizes, they focus their discussion on a binary predictor that assesses whether a virtual address belongs to either a baseline page, or a superpage (in general). We provide an overview about the predictors that they consider.

Figure 8.10 illustrates the two page size predictors considered by prior work [86]. The diagram on the left shows the first approach, a PC-based page size predictor. This approach leverages the fact that the PC of the memory instruction is known well in advance of TLB lookup, so there is enough time available to perform a lookup of a predictor structure before TLB access. The lower-order bits of the PC are used to index a pattern history table (PHT) made up of untagged counters. While a variety of counter organizations are possible (much like branch predictors), the most recent research has studied 2-bit saturating counters. The counters are initially biased toward a strongly predicted baseline page size value. The entry is then trained (on either TLB hits or misses, or both) so that it can record weakly predicted baseline page size, weakly predicted superpage, and strongly predicted superpage. While this approach is simple, a concern is that it learns the page size of a given page separately for each instruction. In other words, multiple predictor entries learn information about the very same page, increasing size requirements.

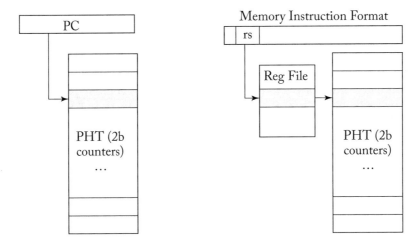

Figure 8.10: The figure on the left shows a PC-based page size predictor, while the figure on the right shows a register-based page size predictor [86].

Consequently, past work has also studied alternate register-based prefetchers, shown on the right in Figure 8.10. Most virtual addresses are calculated using a combination of source register(s) and immediate values. Consider, like prior work, a SPARC architecture [86], where the SPARC ISA uses two source registers (src1 and src2) or a source register (src1) and a 13-bit immediate for virtual address calculation. Prior research shows that the value of the src1 register dominates in the virtual address calculation, often acting as a data structure's base address. Therefore, simply looking up the value stored in the register file in the src1 location can suffice in identifying the desired PHT counter. Furthermore, by using only this base register value and not the entire virtual address, prediction can proceed in parallel with address calculation. This in turn means that TLB lookups are not delayed due to prediction.

Regardless of how prediction is performed, once a guess is made as to whether a virtual page belongs to a baseline page or a superpage, L1 TLB lookups can commence. If, for example, a hash-rehash TLB is used, we first hash using the predicted page size, and rehash with the other page size on a TLB miss.

8.2.3   USING COALESCED APPROACHES

All the approaches to supporting multiple page sizes presented thus far use multiple indexing schemes for different page sizes. As a result, they fundamentally require multiple lookups of the same TLB structure and predictor lookups. An alternate approach, which can simplify the complexity of past approaches is to use a single indexing function for all page sizes. Recently proposed MIX TLBs present such an approach [32].

In order to explain how MIX TLBs operate, consider the address space shown in Figure 8.11. We show virtual and physical address spaces, separating example translations for small pages (A), and superpages (B-C). These addresses are shown in 4 KB frame numbers (full addresses can be constructed by appending 0x000), and are shown for 32 bits to save space. In other words, superpage B is located at virtual address 0x00400000 and physical address 0x00000000. Superpages B and C have 512 4 KB frames in them, indicated by B0-511 and C0-511.

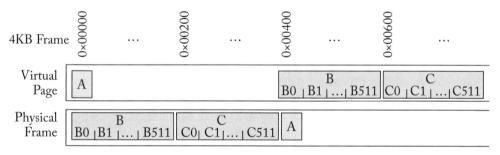

Figure 8.11: Virtual and physical address spaces for an x86-64 architecture, with 4 KB frame numbers shown in hexadecimal [32]. For example, translation B is for a 2 MB page, made up of 4 KB frame numbers B0-B511. 2 MB translations B-C are contiguous.

Figure 8.12 illustrates the lookup and fill operation of MIX TLBs assuming the address space in Figure 8.11, contrasting it with split TLBs. In the first step, B is looked up. However, since B is absent in both the split TLB and MIX TLB, the hardware page table walker is invoked in step 2. The page table walker reads the page table in caches lines. Since a typical cache line is 64 bytes, and translations are 8 bytes, 8 translations (including B and C) are read in the cache line. Split TLBs fill B into the superpage TLB in step 3. Unfortunately, this means that there is no room remaining in the TLB for C, despite the fact that there are 3 unused entries in the

TLB for small pages. MIX TLBs, on the other hand, cache all page sizes. After a miss (step 1) and a page table walk (step 2), we fill B. This necessitates identification of the correct set, presenting a challenge. Since MIX TLBs use the index bits for small pages (in our 2-set TLB example, bit 12) on all translations, the index bits are picked from the superpage's page offset. Thus, superpages do not uniquely map to either set. Instead, we mirror B in both TLB sets.

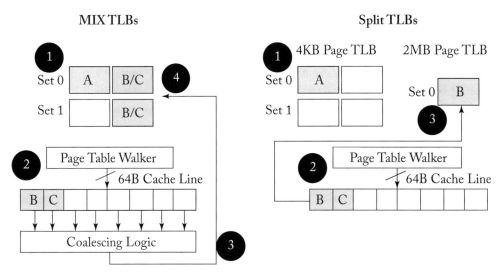

Figure 8.12: Comparing lookup and fill operations for MIX TLBs vs. split TLBs [32].

Naturally, mirroring presents a problem as it uses more TLB entries to store the same information, reducing effective TLB capacity. However, MIX TLBs counteract this issue by observing that OSes frequently (though they don't have to) allocate superpages adjacently in virtual and physical addresses. For example, Figure 8.11 shows that B and C are contiguous, not just in terms of their constituent 4 KB frames (e.g., B0-511 and C0-511) but also in terms of the full superpages themselves. MIX TLBs exploit this contiguity; when page table walkers read a cache line of translations (in step 2) , adjacent translations in the cache line are scanned to detect contiguous superpages. In a manner reminiscent of coalesced TLBs, simple combinational coalescing logic is used for this in step 3. In our example, B and C are contiguous and are hence coalesced and mirrored, counteracting the redundancy problems of mirroring. If there are as many contiguous superpages as there are mirror copies (or close to as many), MIX TLBs coalesce them to achieve a net capacity corresponding to the capacity of superpages, despite mirroring.

The reason that MIX TLBs are appealing is that, as shown in Figure 8.13, lookup is similar to conventional TLBs. While coalesced mirrors of superpages reside in multiple sets, lookups only probe one TLB set. That is, virtual address bit 12 in our example determines whether we are accessing the even- or odd-numbered 4 KB regions within a superpage; therefore accesses to B0, B2, etc., and C0, C2, etc., are routed to set 0. Although we point readers to the original MIX TLB paper for more details, we conclude this subsection by answering some key questions.

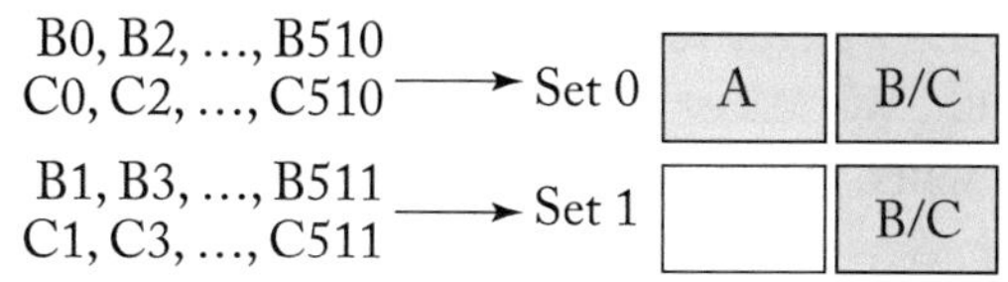

Figure 8.13: Superpages B and C are stored in multiple sets but on TLB lookup, we probe the set corresponding to the 4 KB region within the superpage that the request maps to [32].

Why do MIX TLBs use the index bits corresponding to the small pages? One could, hypothetically, instead consider using the index bits corresponding to the superpage for small pages lookups too. Specifically for our example, we would use virtual address bit 21 as the index assuming we base the index on 2 MB superpages. Conceptually, the advantage of this approach is that each superpage maps uniquely to a set, eliminating the need for mirrors (e.g., B maps to set 0, and C maps to set 1). Unfortunately, this causes a different problem: spatially adjacent small pages map to the same set. For example, if we use the index bits corresponding to a 2 MB superpage (i.e., in our 2-set TLB example, bit 21), groups of 512 adjacent 4 KB virtual pages map to the same set. Since real-world programs have spatial locality, this increases TLB conflicts severely (unless associativity exceeds 512, which is far higher the 4–12 way associativity used today).

Why do MIX TLBs perform well? This is because they are utilized well for any distribution of page sizes. When superpages are scarce, all TLB resources can be used for small pages. When the OS can generate superpages, it usually sufficiently defragments physical memory to allocate superpages adjacently too. MIX TLBs utilize all hardware resources to coalesce these superpages.

How many mirrors can a superpage produce and how much contiguity is needed? The effectiveness of MIX TLBs depends greatly on the available contiguity in superpages, and how much they need to counteract mirroring. Consider superpages with N 4 KB regions, and a MIX TLB with M sets. N is 512 and 262,144 for 2 MB and 1 GB superpages. Commercial L1 and L2 TLBs tend to have 16–128 sets. Therefore, N is greater than M in modern systems, meaning that a superpage has a mirror per set (or N mirrors). If future systems use TLBS where M exceeds N, there would be M mirrors. Ultimately, good MIX TLB utilization requires on superpage contiguity. If the number of contiguous superpages is equal to (or sufficiently near) the number of mirrors , we achieve good performance. On modern 16–128 set TLBs, we desire 16–128 contiguous superpages. Recent work [32] shows that real systems do frequently see this much superpage contiguity.

8.3 TLB SPECULATION

Another approach that is complementary to the idea of increasing the reach of TLBs is to leverage speculation instead, thereby effectively removing expensive page table walks off the critical path of processor execution. Two recent studies [11, 91] have studied TLB speculation in detail and highlighted its benefits. While we point readers to the original papers for the details [11, 91], we provide an overview of the general concept and key design issues here.

The fundamental idea behind TLB speculation is that page tables often exhibit patterns between virtual and physical addresses. If these patterns are repeated amongst translations, and can be learned, it may be possible to speculate the physical page frame number from a virtual page lookup, even when the TLB misses. The processor could then be fed this speculative physical page frame, which could be used to form a speculative physical address that drives a speculative execution path. In parallel, the page table walk can proceed to verify whether speculation was correct. Like all classical speculation, if the page table walker verifies incorrect speculation, the mis-speculated execution path has to be squashed.

In order to speculate effectively, one has to consider what types of virtual-to-physical page patterns are likely to exist in page tables. Prior work on TLB speculation [11, 91] essentially builds on patterns with virtual and physical address contiguity (similar to the work that motivates coalesced TLBs), with two caveats. First, speculative approaches are particularly effective when the contiguity patterns are so large that coalesced TLBs cannot exploit them successfully. Specifically, consider scenarios where several hundreds of contiguous virtual pages are mapped to contiguous physical pages. It is challenging to capture coalesced TLBs using this approach as they rely on bitmaps to record the translations present in a coalesced bundle. If these per-entry bitmaps take up hundreds of bits, TLB sizes become impractically large. Second, past speculation approaches attempt to use the TLB itself (or a similar structure) to perform speculation. That is, the objective is to lookup the TLB, and either achieve a hit, or suffer a miss but speculate immediately on the physical address. However, this requires lookup to remain largely unchanged from standard TLB lookups. For reasons that we will show shortly, prior work [11, 91] achieves this by exploiting contiguity patterns, but only when they are *aligned* to superpage boundaries in the address space.

Figure 8.14 elaborates on these concepts. In the figure on the left, we show a page table with contiguity patterns for translations corresponding to V4–V9. Further, we show another translation mapping V0 to P1. With speculative approaches, we are able to use a single entry in the TLB to map information about V0–V9. Indeed, since prior approaches essentially perform speculation for groups of contiguous 512 pages (aligned 2 MB regions), they can record information about 512 translations within a single TLB entry.

The diagram on the right of Figure 8.14 shows the mechanics of TLB speculation. In our example, like prior work [91], we store speculative entries in the TLB too. One could also use a separate structure [11] to perform speculation but the basic concept remains the same. In this approach, the TLB is first probed. Suppose that the lookup is for virtual page 6. Suppose

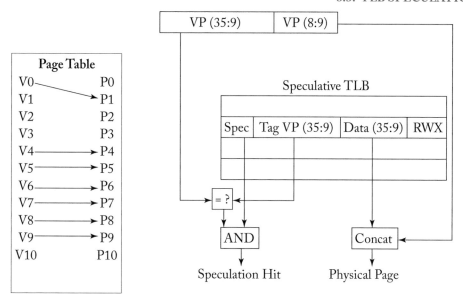

Figure 8.14: On the page table in the left, we would be able to speculate on the contiguous bundle of translations for V4-9. The diagram on the right shows that speculation is performed on 2 MB address boundaries in prior work [11, 91].

further that virtual page 6 has never been accessed in the past, so its entry cannot be resident in the TLB. If, however, virtual page 5 has been accessed in the past, special hardware detects the presence of a large bundle of aligned translations within a 2 MB region, and sets up a speculative entry. Therefore, this speculative entry permits a TLB speculation hit (naturally, the page table walk verifies whether speculation is correct off the critical path). Furthermore, by ensuring that speculation is only permitted for aligned translations in 2 MB regions, constructing the speculated physical address is simply a matter of concatenating bit strings, as shown in Figure 8.14. Finally, note that translations within the same 2 MB region that are not aligned and contiguous (i.e., the translation for V0) can also reside in the TLB, in a separate entry.

The discussion thus far begs the question, how often do patterns like the one presented above, emerge in real systems. Past work has shown two scenarios where such patterns can be prevalent. The first scenario occurs in systems executing OSes that try to generate superpages aggressively. For example, prior work [11] has shown that OSes like FreeBSD, which use reservation-based superpaging, often see instances where hundreds of translations for 4 KB regions are lined up inside a 2 MB region in anticipation of promoting them to a single superpage. However, because the application may need to use a certain amount of memory before this promotion occurs, or because of the presence of IO pages, these 2 MB regions often remain splintered into several hundreds of 4 KB contiguous pages for long periods of time. In these situations, speculative TLBs can successfully use a handful of TLB entries to capture information about hundreds of such translations. In effect, the bulk of page table walks are removed from the critical path of execution. Similarly, another source of massive amounts of aligned contiguity arises from situations where the OS can form superpages, but then has to break them up or splinter them for overall performance reasons [91]. This can occur for several reasons including a need for finer-grained memory protection, to lower paging traffic, and to lower overall memory consumption. As with reservation-based superpage creation, the process of breaking superpages also presents effective opportunities for TLB speculation.

8.4   TRANSLATION-TRIGGERED PREFETCHING

All the optimizations we have discussed thus far reduce the frequency of TLB misses, or remove TLB miss penalties from the critical path of execution, mitigating their performance impact. However, TLB lookup and miss handling represent only one portion of the address translation stack. The other, largely overlooked component is the memory replay that follows the virtual-to-physical page translation. We define replay as follows. Suppose that the CPU generates a memory load to a virtual address on the third cache line in virtual page 2. As per normal, the CPU aims to translate this virtual address to a physical address. There are two possibilities. In the first case, the translation is found in the TLB, leading to a TLB hit. In the second case, the translation is absent and there is a TLB miss. TLB misses are serviced by hardware walkers which identify the desired translation from the software page table residing in the memory hierarchy. Subsequently, the originally desired memory reference must itself complete. Suppose virtual page 2 maps to physical frame 6. This means that ultimately, there is a memory replay for the third cache line in physical frame 6.

Suppose that we focus on the situation where the CPU has a TLB miss, which is serviced by the page table walker, before it performs a replay. One might, at first blush, expect the TLB miss to be more harmful than the replay in terms of performance, since a page table walk requires multiple sequential memory references to be serviced (four for x86-64 systems) while a replay requires one memory reference. In reality, however, recent work shows that replays in real-world workloads are often just as harmful to performance as TLB misses [19]. The main reason for this is that processor vendors have invested years of research and development on hardware acceleration of TLB misses. Innovations like MMU caches and multiple page table walkers (see previous chapters) mean that TLB miss penalties now mostly consist of the cost of a single memory reference for the leaf page table entry, since higher level entries from the page table are generally (90%+ of the time) found in the MMU caches. Therefore, the bulk of TLB miss overheads end up being composed of the one reference for the leaf page table entry, and one for the replay [19].

This begs the question: beyond accelerating TLB misses, can we also accelerate the replay itself? Recent research shows how to achieve this by cache prefetch of the data required by the replay, triggered by the page table walk [19]. The basic insight is this. Suppose that the CPU experiences a TLB miss, triggering a page table walk. Suppose further, that the lookup for the leaf page table entry results in L1-LLC cache misses, and requires a lookup of main memory. Recent work shows that if page table entries are poorly cached, the physical page that they point to is likely to be poorly cached too [19]. This is intuitive since page table entries must generally be accessed on the path of accessing a cache line from the physical page that the page table entry points to. This presents an opportunity: when main memory must be accessed to satisfy a page table lookup, can we deduce the physical memory address that the subsequent replay will need and prefetch it into the LLC to enable better performance?

Figure 8.15 graphically illustrates translation-enabled memory prefetching optimizations or TEMPO [19] to accelerate memory replays. The x-axis plots the latency of the page table access, while the y-axis plots the latency of the replay. TEMPO is triggered when DRAM needs to be accessed for the page table lookup. By prefetching data into the LLC, access times for replays are reduced to the latency of an LLC hit, accelerating performance. While prior work has not considered prefetching into higher levels of the cache (e.g., even the L1 cache), such studies are likely ripe for future exploration. We now focus on two issues fundamental to TEMPO's operation. First, how can we deduce, when performing main memory lookup for the page table entry, the target physical address of the replay? Second, how do we ensure prefetching timeliness?

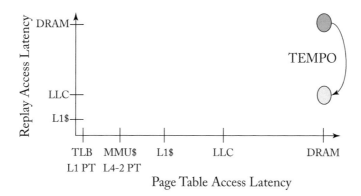

Figure 8.15: Translation-enabled prefetching optimizations (TEMPO) are used to accelerate the penalty of the memory replay after a TLB miss [19].

Constructing the target physical address of the replay: Figures 8.16–8.18 show how TEMPO constructs the target replay physical address to enable prefetching. Since prefetching is triggered on a memory access, the initial work on TEMPO [19] adds logic in the memory controller (MC) hardware to calculate the replay's physical address. In the example of Figure 8.16, the target replay address is 0x3002, meaning that the first cache line in physical page 0x3 must be accessed. Consequently the MC needs to make a request for address 0x3000. This opens up the question of the MC deduces that 0x3000 is the target address.

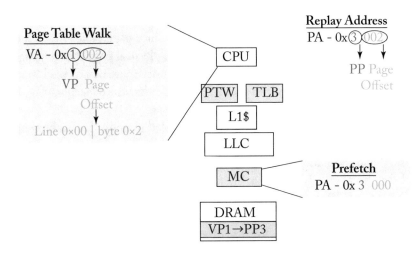

Figure 8.16: **TEMPO** requires that the memory controller perform a prefetch triggered by main memory access for a page table entry [19].

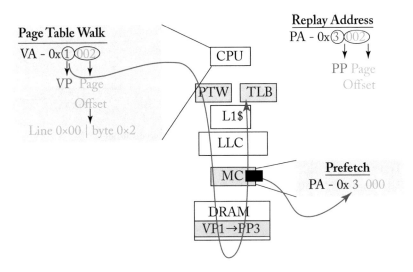

Figure 8.17: We add logic in the memory controller to extract the physical page frame number bits from the page table entry triggering the prefetch [19].

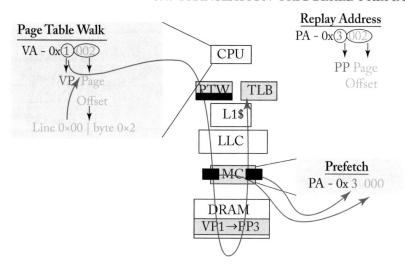

Figure 8.18: TEMPO requires the page table walker to append the message that it normally sends out with the cache line number within the page that is requested [19].

Figure 8.17 shows the first step in calculating this target address. Since the prefetch trigger is a lookup for a page table entry, the MC has to initially lookup the desired translation (V1 mapping to P3 in our example). This presents the following opportunity: since the translation maintains information about the replay target's physical address (0x3), we can add logic in the MC to extract these bits from the translation that is looked up. Doing this requires simple combinational hardware additions to conventional MC logic, and fulfills the calculation of several of the replay target's address bits.

Figure 8.18 shows the second step in replay target address calculation. The MC needs to infer the bits corresponding to the first byte of the cache line number where the replay target resides. To percolate these bits to the MC, the original TEMPO work [19] augments the messages sent by the hardware page table walker to the memory hierarchy with the bits representing the cache line number. This constitutes an addition of 9 bits to the messages normally sent by the page table walker. Ultimately, this (slightly) higher traffic is offset by TEMPO's performance gains.

Performing timely prefetching: Having described the process of calculating the replay target address, we now discuss timeliness issues. Naturally, in order for prefetching to be useful, we want to conclude the prefetch of the replay target data into the LLC prior to the replay's LLC lookup. Figures 8.19 and 8.20 compare these scenarios. By prefetching at DRAM access for the page table entry, most modern systems provide a window of roughly 100-150 clock cycles for the replay target prefetch to be filled into the LLC. While prior work has more details on this [19], this timescale is usually enough for LLC prefetch.

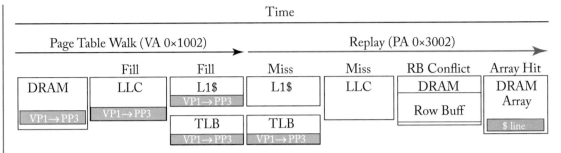

Figure 8.19: Timeline of events corresponding to a TLB miss and replay [19].

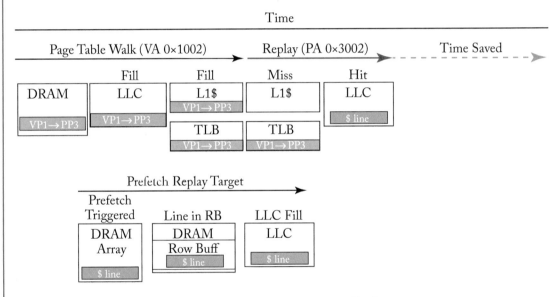

Figure 8.20: Timely TEMPO prefetching saves replay time [19].

8.5   OTHER IMPORTANT HARDWARE IMPROVEMENTS FOR VIRTUAL MEMORY

We have focused, in this chapter, on hardware mechanisms to improve address translation performance. Our studies are, however, non-exhaustive. In particular, we point readers to research on an emerging area of increasing importance in VM design: the notion of translation coherence. In particular, with the emergence of heterogeneous memory architectures, it is becoming increasingly important to move pages of memory between devices with complementary performance characteristics. For example, it may be beneficial to migrate pages of memory from DRAM devices with high capacity to higher-bandwidth but smaller die-stacked DRAM. In these situations, page tables must be changed and TLB coherence can become a performance bottleneck. We point readers to recent work on adding instruction support and hardware to improve TLB coherence [110], as well as schemes that improve translation coherence by fusing them with existing cache coherence protocols [95, 113].

8.6   SUMMARY

This chapter's focus was on emerging research questions on VM, particularly in the context of hardware innovations. We described several recent research studies, some of which have influenced real-world products (e.g., coalesced TLBs), to improve TLB capacity, miss penalties, and page table walks. Several of these proposals remain in nascent stages of development and therefore warrant further studies. We encourage interested researchers to further delve into these problems, particularly as different approaches are likely necessary for alternative computing units like GPUs, and accelerators.

CHAPTER 9

Advanced VM Hardware-software Co-design

The last chapter presented recent research efforts targeting more efficient address translation, but focused on hardware optimizations. In this chapter, we study techniques requiring hardware-software co-design. Like the previous chapter, the following discussion presents a non-exhaustive list of recent research. In fact, there is an interesting body of recent work that focuses on purely software optimizations to improve VM performance. While this work is certainly relevant to graduate students exploring this area, it requires detailed discussion of core operating system design and implementation issues, beyond the scope of this synthesis lecture. Nevertheless, we briefly point students to two general streams of recent work on purely software topics:

OS support for transparent superpages: Recent years have seen many studies on OS support for large pages or superpages. The traditional approach for superpage support had been through the use of a dedicated library (e.g., *libhugetlbfs* for Linux) that programmers had to explicitly link to. Among many programmability challenges, a key issue with this approach is that programmers need to know exactly how many superpages their programs are likely to need, and which data structures are likely to benefit from superpage allocation. This can be challenging to decipher, as several studies have recently shown [8, 45, 91]. Consequently, software developers and researchers have been studying better approaches for *transparent* superpage allocation, where the programmer is relieved of the burden of identifying when superpages are useful, with the OS taking on this responsibility instead. In this approach, the OS opportunistically attempts to create superpages in a manner that is transparent to the program. We point interested readers to the Linux community's support for transparent superpages [8], which has focused on 2 MB transparent superpages. More recently, the Linux community has also begun investigating support for 1 GB transparent superpages too. Additionally, recent academic studies on Ingens [74] show how transparent superpage support can be further improved by managing contiguity as a first-class resource. In so doing, Ingens tracks page utilization and access frequency, eliminating many fairness and performance pathologies that otherwise plague modern OS support for superpages.

OS support for efficient TLB coherence: As discussed at the end of the last chapter, the systems community has also recently been studying ways to improve the overheads of TLB coherence. These studies have been driven by several technology trends, especially that of increasing system heterogeneity. Many recent studies aim to improve TLB coherence performance in hardware [89, 95, 110, 113], but several recent studies also consider software solutions that can be readily deployed on real systems. We point readers to two recent studies on this. The first proposes mechanisms to patch TLB coherence code into the microcode of x86 cores, thereby avoiding many of the performance overheads of TLB shootdowns [85]. The second approach [7] optimizes TLB coherence operations by better detecting which cores cache specific page table entries in their TLBs. Both these studies essentially present ways of scaling core counts on systems without crippling them with TLB coherence operations.

Having briefly discussed—and pointed out to interested readers—relevant work on software-only techniques to improve VM, we now turn our attention to three recent studies on hardware-software co-design. All three studies essentially aim to boost TLB reach by adding intelligence to the OS VM stack, and modestly augmenting the TLB to take advantage of these changes. We begin with such a study on TLB prefetching, before turning our attention to alternative ideas.

9.1 RECENCY-BASED TLB PRELOADING

Like caches, TLBs are amenable to prefetching. Most prior studies on TLB prefetching have focused on predicting which translations are likely to be accessed in the near future after a TLB miss. Subsequently, these translations are fetched by the page table walker in parallel with CPU execution, with the goal of filling them in the TLB in a timely manner right before they are demanded by the CPU. TLB prefetching work has generally focused on three methods of predicting which translations are likely to be accessed in the near future.

The first method is analogous to stride prefetching techniques already used for caches. In this approach [65], TLBs are augmented with stride detection hardware. When this hardware detects repeated strides in access patterns (e.g., if subsequent loads tend to access addresses residing in virtual pages that are N pages apart), it calculates the virtual page numbers of future accesses. The page table walker is then tasked with looking up the page table for these virtual pages, and prefetching the corresponding translations into the TLB.

The second method goes beyond distance-based prefetching approaches for uniprocessors and discusses prefetching techniques for multi-core systems [20, 22]. A key observation with this approach is that multiple threads of parallel programs running on separate cores often share data structures and hence require the same sets of translations in their TLBs. In response, recent work proposes hardware TLB prefetching schemes where cores study the TLB access patterns of one another to prefetch translations.

While these approaches are effective, they require additional hardware to predict future translation targets. In contrast, a third method, which is the subject of this section, called recency-based TLB preloading modifies the page table modestly to enable predictions of future accesses translations. The hardware page table walker is augmented (in a very simple manner) to read the additional page table metadata that dictates where to prefetch from, and to then perform the prefetches using standard page table walks. We now delve into recency-based TLB preloading.

Basic idea: Recency-based preloading is based on the notion of recency of use of translations [97]. Recency of use is a well-known concept from stack algorithms and can be understood as follows. Suppose we use a logical stack of translations to denote which translations have been accessed most recently. As a translation is accessed, it is placed at the top of the stack. Previously accessed translations are maintained below the top of the stack, in LRU order.

This recency-based stack can be used to assess TLB miss rates for any fully associative LRU TLB size. To see how, consider an infinite LRU stack. We examine each memory reference in the order presented to the TLB. If the reference's address has been used before, it resides at level R in the stack. This is known as the recency of the reference. If R is less than N, the size of the TLB whose miss rate we are trying to determine, the reference enjoys a TLB hit. This entry is then removed and pushed on the top of the stack. If, however, there is no match, the address reference is pushed on the top of the stack and the entry at level N becomes the replacement victim.

Using this stack algorithm, we can determine the miss rates of all possible TLB sizes. In fact, the authors of the original study [97] do this to count the number of times an address is found at a particular recency or depth in the stack. In so doing, they observe the following: accesses to a translation residing at a certain stack depth are usually immediately followed by accesses to a similar depth.

Figure 9.1 illustrates this concept. Suppose that at the beginning of our memory references, our program has accessed virtual pages A, B, C, D, P, and L. Based on the recency stack, D has been accessed most recently while L has been accessed furthest back in time. Suppose that at time t0, there is a reference for virtual page C. There is a hit at a recency or stack depth of 3. C is therefore brought to the top of the stack. In the next time epoch, t1, programs often tend to access memory at a similar stack depth, in this case to page A again at a stack depth of 3. Successive accesses are to stack depths of 4 and 2, which are again similar to depths of 3.

The observation that there is temporal locality in access of stack depth is the crucial insight behind recency-based preloading. In order to see how it works, let us consider, without loss of generality, a fully associative TLB with LRU. Naturally, this type of TLB constitutes the upper portion of the recency stack. That is, consider an N-entry TLB. In this case, we can think of the recency stack as having its first N entries devoted to the TLB, while the remaining entries reside in main memory and hardware caches. Figure 9.2 shows an example of such a case, where we maintain a 4-entry TLB to cache translations for A–D. Suppose that the CPU accesses a VM address located in virtual page P in time t0. Consequently, the translation for P is unhooked from the recency stack and brought to its head, which exists in the TLB. In order to make room for this entry, the translation residing at the LRU position in the TLB, translation B, is evicted

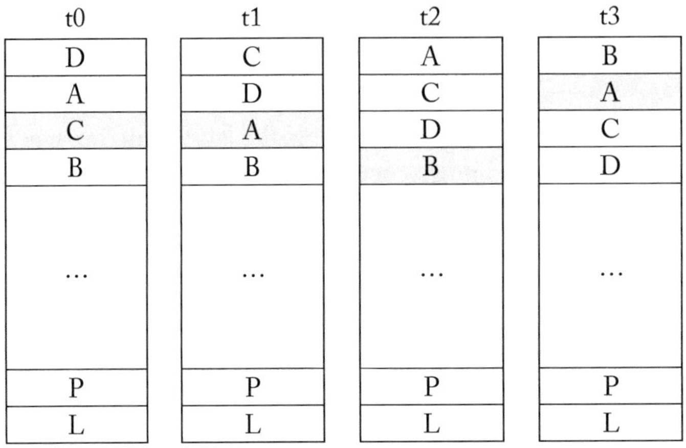

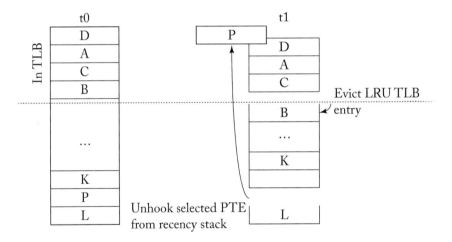

Figure 9.1: Successive accesses tend to be to the same stack depth in the recency stack [97].

Figure 9.2: TLB misses bring in an entry from a certain stack depth, evicting an entry to the head of the recency stack portion maintained in memory [97].

from the TLB. This means that it is now inserted at the head position of the portion of the recency stack that is non-TLB resident.

When P is filled into the TLB, recency-based preloading also performs the following operation. It anticipates that the CPU will also need translations residing at similar stack depths in the near-future. Dedicated logic in the hardware page table walker identifies that translations K and L are at similar stack depths, and should therefore also be prefetched into the TLB.

Although not shown in Figure 9.2, these two translations are subsequently prefetched into the TLB.

Implementation: The critical question in the design of recency-based TLB preloading is how the recency stack should be implemented. Since the recency stack is implemented in two separate areas, the TLB and the page table, we separately discuss each case.

The original recency-based preloading paper [97] discusses TLB implementation issues. A fully associative LRU TLB naturally maintains the TLB's portion of the recency stack. However, we have discussed that most real systems do not maintain fully associative TLBs because of their energy and power overheads. However, prior work discusses that modern L2 TLBs with relatively high set-associativities, approximate a recency stack in any case.

Implementing a recency stack in the page table is more complex. Figure 9.3 shows how it is maintained. The change to the page table is that each translation entry now houses two additional fields. These fields maintain pointers that track the translation numbers at a recency depth of +1 and −1, corresponding to a forward and backward pointer respectively. The page table walker is responsible for setting up these pointers and reading them as follows.

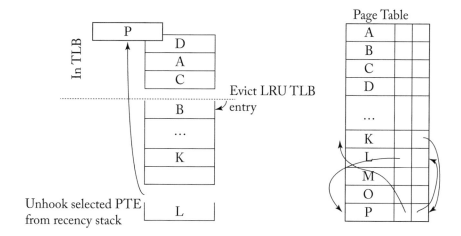

Figure 9.3: The recency stack is maintained in the page table by adding forward and backward pointers to each translation entry [97].

Suppose that we experience a TLB miss for the translation for virtual page P. When P is filled into the TLB, the forward and backward pointer fields of P's translations are read by the hardware page table walker. As shown in Figure 9.3, the forward-pointer points to K, while the backward pointer points to L. Therefore, the basic recency-based preloading scheme prefetches K and L into the TLB too. Although we just illustrate prefetching translations +1 and −1 in the recency stack, one could also follow chains of pointers and prefetch translations at positions of 2, 3, etc., in the recency stack. In the next step, B is evicted from the TLB. Consequently, the hardware page table walker sets B up at the position of the top of the recency stack in the page table. We point interested readers to the original paper [97] for more details.

The key innovation with recency-based preloading is that it constructs the first instance of a TLB prefetcher that goes beyond stride prefetching. While it has yet to be adopted by industry, we believe it presents an interesting design point worthy of further investigation, especially in the context of emerging hardware accelerators.

9.2   NON-CONTIGUOUS SUPERPAGES

In the previous sections, we discussed the opportunities and challenges imposed by superpages on address translation performance. While superpages, when used judiciously, promise good performance, emerging technology trends pose problems for their continued use. In particular, recent studies show that permanent DRAM faults are becoming more common, and cause pages of physical memory to be retired from use by the OS [40]. This presents a challenge for OSes trying to construct superpages, especially when they are large (e.g., 1 GB superpages) since just a few retired 4 KB regions in physical memory can preclude superpages almost entirely. The main problem is this: superpages require large swathes of contiguity in virtual and physical space. For example, 2 MB superpages require aligned contiguity of 512 4 KB physical address spaces. If even a single 4 KB physical frame has to be retired from this contiguous range, the superpage cannot be realized.

While the issue of memory faults can be partly handled by TLB coalescing, recent work studies more flexible changes to the page table structure to accommodate these changes [40]. This new approach to maintaining page tables is known as gap-tolerant sequential mapping (GTSM). Figure 9.4 shows an example GTSM's operation, and contrasts it against traditional superpages and baseline page sizes.

The page tables on the left in Figure 9.4 show an example of a mapped superpage in green. As per usual, a large aligned and contiguous bundle of virtual address space is mapped to a corresponding space in the physical address space. The diagram in the center of Figure 9.4, in comparison, shows a physical address space where three physical pages have to be retired because of memory faults (shown in black). As a result, superpages cannot easily be formed and we are left with smaller baseline pages. The diagram on the right in Figure 9.4 shows how GTSM overcomes this problem.

With GTSM, a superpage in the virtual address space is divided into smaller fixed-sized virtual address blocks, which are sequentially mapped to *building-blocks* in the physical address space. Figure 9.4 shows the building-blocks in blue. Building-blocks are bigger than the regular page but are smaller than superpages. Further, a superpage in virtual address space can map to non-contiguous building-blocks in physical address space, as long as those building-blocks reside in a portion of the physical address space that is limited to twice the size of a superpage. This region of physical address space is called a *memory slice*. As detailed in the original paper [40], GTSM is therefore a generalized form of traditional superpage mapping, but is more

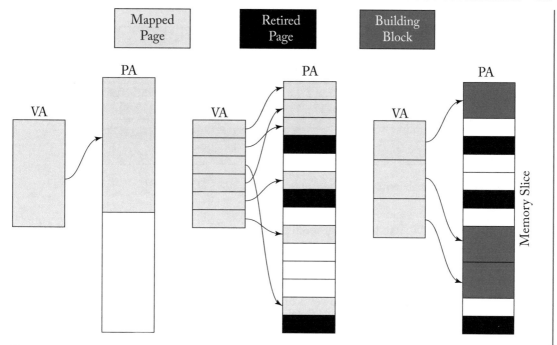

Figure 9.4: We form non-contiguous superpages my considering memory slices that contain building blocks of contiguous page frames [40].

flexible in that it supports page retirement. In the absence of page retirement, GTSM creates the same memory mapping as traditional superpages. Finally, exactly half of the building-blocks in a memory slice participate in a GTSM mapping. Any building-block that contains at least one retired region cannot be used in the mapping.

To implement GTSM, the authors of the original study modify the basic structure of the page table [40]. To demonstrate the necessary changes, consider an x86-64 page table table. The authors change the page directory entry level, or the L2 level, which records information about 2 MB superpages. To support GTSM, the page table entry at this level, which is 8 bytes, is extended to be 16 bytes. In other words, a single GTSM entry is formed by using two adjacent L2 page table entries. Figure 9.5 shows the resulting page table entry. At the top, we show the building-blocks in a memory slice (for our example, we show 64 building-blocks). The shaded blocks are allocated as part of a GTSM superpage. The diagram at the bottom of Figure 9.5 shows that fusing two original L2 page table entries together permits using one of them to maintain a bitmap. This bitmap indicates which of the building-blocks in the memory slice constitute this GTSM superpage. Note that only half the bits in the bitmap can be set at any given point, since a GTSM superpage can accommodate half the building-blocks in a memory slice.

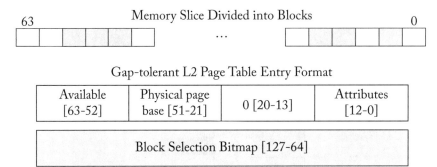

Figure 9.5: L2 page table entry modified to accommodate GTSM [40].

Figures 9.6 and 9.7 contrast traditional superpage address translation and GTSM address translation. As shown in Figure 9.7, a GTSM mapping essentially maps a 4 MB portion of the physical address space. Only the upper 8 bits of the L2 index are needed to index the GTSM page table page, since there are up to 256 GTSM page table entries in it (see the original paper for details [40]). Therefore, only the lower 5 bits of the L2 index are used as block offset and are kept unchanged during address translation. The remaining 5 bits are treated as an index into the block selection bitmap. Finally, because the mapped sliced is aligned at an 8 MB boundary, the low 2 bits of the physical page base address field are always zeros and ignored in the translated physical address. To translate block index K (0-31), the block selection bitmap is scanned to find the K selected bit, whose position in the bitmap (0–63) indicates the B-block that the virtual block is mapped to.

While GTSM primarily requires changes to the OSes page table structure, it also requires modifications to the hardware page table walker and to the hardware MMU cache that maintains page table entries from the L2 page table page [10, 17, 18]. Changes to the page table walker are straightforward in that its finite state machine needs to be able to construct the lookup shown in Figure 9.7. Changes to the L2 MMU cache are, however, more invasive since each entry must now be also accommodate the block bitmap. This increases the L2 level MMU cache. Nevertheless, the original study shows that this trade-off ultimately provides good performance in situations where memory pages are retired sufficiently to fragment traditional superpages [40].

9.3 DIRECT SEGMENTS

The final advanced topic we cover in this chapter pertains to the concept of direct segments [12], which combines several desirable attributes of paging and the segmentation approaches discussed at the beginning of this synthesis lecture. The original direct segments paper [12] asks the following question: what aspects of VM are actually used by big-memory workloads today? The workloads profiled by the authors include memory-intensive workloads such as databases, key-value stores, graph algorithms, as well as high-performance computing applications with

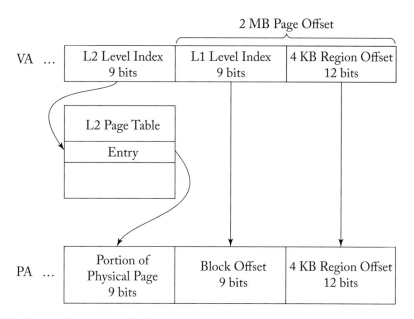

Figure 9.6: Traditional superpage lookup.

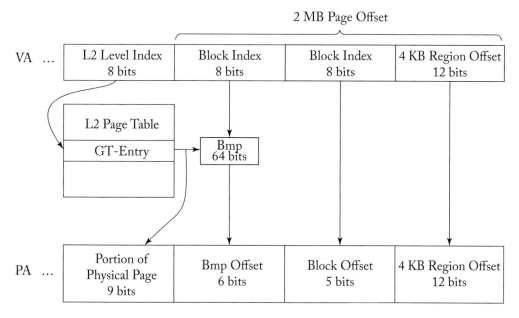

Figure 9.7: GTSM lookup and physical address construction [40].

large memory requirements. Based on an exhaustive set of studies the authors conclude that for the vast majority of the program's address space, the workloads do not require swapping, fragmentation mitigation, or fine-grained protection afforded by current VM implementations. In fact most of these workloads allocate memory early and have stable memory usage. In particular, the authors study the following:

Swapping: One of the primary reasons for the success of page-based VM was that it automatically managed scarce physical memory without explicit programmer management. In particular, the OS would swap pages between memory and backing storage to provide programmers the illusion of a vast memory space, much bigger than the amount of actual physical memory in the system.

However, the direct segments paper shows that modern big-memory workloads do little or no swapping. The reason is that many of these applications are performance-critical and cannot afford to wait for disk access. For example, Google observes that sub-second latency increases can reduce user traffic by as much as 20% [12]. Hence, like Google, Facebook, Microsoft, and Twitter keep user-facing data like search indices in main memory. Similarly, enterprise databases and in-memory object caches exploit buffer pool memory to minimize disk access. Overall, many real-world systems are provisioned with sufficient physical memory for the entire dataset or a large fraction of it, largely reducing the need for much swapping.

Memory allocation and fragmentation: Another aspect of big-memory workloads is that they tend to allocate most of their memory during startup and manage that memory internally. For example, databases like MySQL allocate buffer pool memory at startup and then use it as a cache, query scratchpad, etc. Memcached similarly allocates space for its in-memory object cache during startup. Consequently, most big-memory workloads see little variation in allocated memory after the workloads begins execution. Furthermore, since many of these workloads are long-running, the actual allocation phase is amortized over the runtime of the application.

Per-page permissions: Finally, the workloads evaluated in the original direct segments paper [12] also see many scenarios where vast swathes of pages share the same permission attributes. For example, many of the workloads dynamically allocate memory at startup with read-write permission. At the same time, however, there are situations when finer-grained memory protection become necessary. For example, memory regions used for inter-process communication use page-grain protection to share data/code between processes. Code regions are protected by per-page protection to avoid overwrite. Copy-on-write uses page grain protection for efficient implementation of the fork() system call to lazily allocate memory when a page is modified.

At a high-level, these observations suggest that there are a class of workloads which pay the performance penalty of looking up a TLB for small page sizes, but do not actually use the flexibility of fine-grained paging. Naturally, this observation is true only for the specific big-memory workloads that the authors consider. Nevertheless, these big-memory workloads are in widespread use and warrant more efficient VM support. Ultimately, the authors conclude that modern systems must continue supporting all the features of VM that are classically important (i.e., swapping, fine-grained protection, support for dynamic allocation and defragmentation), but that there should also be parallel hardware/software support or "fast-paths" for the types of big-memory workloads that they authors consider.

The notion of direct segments is essentially a realization of this observation. A direct segment maps a large range of contiguous VM addresses to contiguous physical memory addresses using small, fixed hardware: base, limit, and offset registers for each core. If a virtual address is between the base and limit, it is translated to a physical address with the corresponding offset within the direct segment. This obviates the need for a TLB miss. The original direct segments paper [12] stipulates that all addresses within the segment must use the same access permissions and reside in physical memory. Furthermore, direct segment support operates harmoniously with situations/workloads which continue to require conventional page-based VM. In fact, virtual addresses outside the segment's range are translated through conventional page-based VM using TLBs and hardware page table walkers. Direct segments are expected to be used to map the large amounts of VM typically needed by big-memory workloads at startup. We now present details of the hardware and software support required for direct segments.

9.3.1   HARDWARE SUPPORT

With direct segments, a program's memory reference has two execution paths. The first one is via the conventional TLB hierarchy and follows conventional paging. The second targets memory references within the direct segment, and accesses the contiguous physical address range through hardware support in a manner that precludes TLB misses. This contiguous virtual address range can be arbitrarily large and is limited only by the physical memory capacity of the system. In other words, a program can opt to map all of physical memory with a direct segment, can opt to map all of physical memory using conventional paging, or can choose a combination of the two. Such flexibility ensures that background processes are unaffected and backward compatibility is maintained.

Figure 9.8 shows how direct segment registers are used to calculate physical addresses residing in the contiguous physical address range. Each core maintains, in addition to the standard TLB and page table walker, several registers: (1) a BASE register holding the starting address of the contiguous virtual address range mapped through the direct segment; (2) a LIMIT register holding the end address of the virtual address range mapped through the direct segment; and (3) an offset holding the start address of the direct segment's backing contiguous physical memory minus the value in the BASE register.

By design, direct segments are aligned to the base page size, so page offset bits are omitted from these registers. As shown, suppose that there is a memory reference to an address V. This is split into a virtual page number and a page offset. The page number is compared to the BASE and LIMIT register, in parallel with the standard TLB lookup. If the page is found to reside between these values, we can identify that the address resides in the direct segment. In this case,

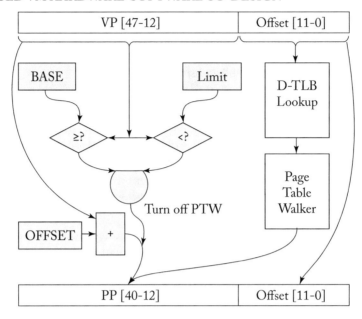

Figure 9.8: Direct segment registers are used to construct the physical page number [12].

a signal is routed to the page table walker to prevent it from performing a page table lookup (since the TLB must have experienced a miss) and an offset register is added to the location of the virtual address within the direct segment to construct the desired physical page number.

Note that direct segments permit read-write access only, although additional hardware could be added to permit more flexible use of permission bits. Further, the OS is responsible for loading proper register values, which are accessible only in privileged mode. Setting LIMIT equal to BASE disables the direct segment and causes all memory accesses for the current process to be translated with paging. Finally, the original direct segments paper emphasizes three important points.

1. Direct segments do not export two-dimensional address space to applications, but retain a standard linear address space.

2. They do not replace paging as addresses outside the segment use paging.

3. They do not operate on top of paging. In other words, direct segments are not paged.

9.3.2   SOFTWARE SUPPORT

The primary job of the OS is to provide an abstraction for programmers to allocate data structures with a direct segment. To do this, the OS separates each application's address space into two portions. One is a *primary region* where the direct segment can be allocated. The remainder of the address space is the normal pageable and swappable address space we use conventionally.

A primary region is a contiguous set of virtual addresses with read-write permissions. It maintains no guarantees about paging, fine-grained protection, swapping, etc. In other words, it is ideal for the majority of the big memory workloads' memory usage, e.g., mySQL's buffer cache or memcached's cache. Software must provide two types of support for this primary region: (1) the ability to provision a range of VM addresses as a primary region; and (2) enable memory requests from an application to be mapped to it.

Consider the notion of provisioning a range of VM addresses as the primary region. During the creation of a process, the OS can reserve a contiguous address range in its address space, to be used for memory allocations in the primary region going ahead. This partition must be big enough to encompass the largest possible primary region, which is usually the size of the physical memory. Given the vast address range afforded by 64-bit address spaces, there is plenty of room in the virtual address space to reserve space for the entire primary region.

One provisioning is handled, the OS must decide which memory allocation requests to use the primary region for. Two approaches, opt-in or opt-out, are possible. A process may explicitly request that memory allocation be put in its primary region via a flag to the memory allocator (e.g., mmap). Alternately, a process may default to placing dynamically allocated non-file-backed memory with read-write permissions into the primary region. Anonymous memory allocations can include an opt-out flag when paging features are needed, such as sparse mapping of virtual addresses to physical memory.

The original direct segments paper [12] covers additional details on how the base, limit, and offset registers are set up and maintained by the OS. While we refer interested reader to this paper for more details, the general idea essentially boils down to the observation that it may be prudent to consider multiple approaches to VM. In this specific case, direct segments are permitted when the programmer understands that his/her programmer is amenable to it, and paging is used otherwise.

9.4   OTHER HARDWARE-SOFTWARE APPROACHES

The direct segment approach has inspired a number of related studies that we point readers to. For example, the direct segments approach has also been applied to virtualization to essentially short-circuit the overheads of two-dimensional page table walks. Specifically, recent work [43] shows that it is possible to use direct segments at the host or guest level, or both, almost entirely eliminating the TLB overheads of virtualization in many big-memory workloads.

Yet another interesting avenue is recent work on range translations [66]. This line of work builds upon direct segments and considers the implementation challenges of supporting multiple arbitrarily sized ranges of contiguous virtual and physical addresses. In other words, it replaces a single direct segment with the ability to create multiple smaller ranges, thereby improving the performance of even more workloads than the original direct segments work. We point readers to the original paper for more details. Furthermore, we encourage readers to investigate the use of concepts like direct segments and range translations in emerging accelerator architectures, for which novel address translation hardware is a first-class design goal.

9.5 SUMMARY

This chapter went beyond prior work and discussed the potential for better hardware-software co-design of the VM subsystem. Many of the proposed approaches discussed ways to alleviate TLB misses, either by intelligent software-directed prefetching, or a better understanding of what aspects of VM are actually needed by modern workloads. These observations are tightly tied to the particular workloads that we focus on. It will be interesting to explore how emerging workloads targeted at domains like deep learning, virtual/augmented reality, and so on continue to stress the VM system in new and interesting ways. The field of computing is a fast-moving world, and since VM technology powers most non-trivial computer architectures today, we expect the VM paradigm to continue evolving and to remain a deep and open area deserving continued attention and research well into the foreseeable future.

CHAPTER 10

Conclusion

This synthesis lecture explored the classic computer science abstraction of VM. Virtual memory is a decades-old concept that is fundamental to the programmability, portability, and security of modern computing systems of all scales, ranging from wearable devices to server systems for warehouse-scale computing. Indeed, a measure of virtual memory's success is that programmers rarely think about it when writing code today. As computer systems accommodate new classes of software, and integrate specialized hardware and emerging memory technologies, it is vital that we preserve and rethink the VM abstraction to ensure that these systems remain programmable. As we have discussed, however, these hardware and software trends also stress our current implementations of VM. As such, one of the important puzzles facing the system community is how to redesign the concept of VM in a computing landscape that is different from the era of mainframes with discrete electronic components, when VM was first conceived.

This book attacks this problem by covering the fundamentals of VM and also recently proposed techniques to mitigate the problems facing it today. One class of techniques that we cover consists of hardware-based approaches (e.g., shared TLBs, coalesced TLBs, part-of-memory TLBs, etc.). The benefit of hardware techniques is that they do not require OS or application-level changes. Consequently, if the hardware remains modest in implementation requirements, it may be more feasible for integration into full systems today. On the other hand, hardware-software co-design (e.g., direct segments, etc.) present the potential to dramatically reduce address translation overheads. The caveat is that more layers of VM require change.

While these studies present a start, a range of important and fundamental questions remain unaddressed. As just one example, the notion of a page as the basic unit of allocation, hardware protection, and transfer between memory and to secondary storage opens up lots of questions. With emerging memory technologies like byte-addressable non-volatile memory, what should the size of the page be? The "right" size is based on a variety of factors like memory and disk fragmentation, amortizing the latency of disk seeks, and minimizing the overhead of page table structures. These tradeoffs change with newer memory technologies. Similarly, a range of questions that explore the interactions between filesystem protection and memory protection, the role of superpages and their relationship to not just address translation but also memory controllers [45], etc., remain to be explored.

We end this book by reiterating a theme that we have addressed several times in this lecture. The VM subsystem is a complex one, and requires careful coordination between the hardware, operating system kernel, memory allocators, and runtime systems/libraries. Consequently, VM layers have historically been the source of several high-profile bugs at the hardware and software layers. As we augment existing hardware and software, and propose more radical changes to VM, it is important that we consider the verification challenges posed by these changes. We therefore believe that as systems continue to embrace complexity, it will ultimately be necessary to carefully model the impact of VM innovations on the full computing stack, from the OS level down to the register-transfer. We believe that automated approaches to achieving this therefore remain a fruitful research direction, along with more "traditional" approaches that seek to optimize performance and energy.

Bibliography

[1] Neha Agarwal, David Nellans, Eiman Ebrahimi, Thomas F. Wenisch, John Danskin, and Stephen W. Keckler. Selective GPU caches to eliminate CPU–GPU HW cache coherence. In *IEEE International Symposium on High Performance Computer Architecture (HPCA)*, 2016. DOI: 10.1109/hpca.2016.7446089. 87

[2] Neha Agarwal, David Nellans, Mark Stephenson, Mike O'Connor, and Stephen Keckler. Page placement strategies for GPUs within heterogeneous memory systems. *International Conference on Architectural Support for Programming Languages and Operating Systems*, 2015. DOI: 10.1145/2694344.2694381. 43

[3] Neha Agarwal and Thomas Wenisch. Thermostat: Keeping your DRAM hot and NVRAM cool. *International Conference on Architectural Support for Programming Languages and Operating Systems*, 2017. 43

[4] Alfred Aho, Peter Denning, and Jeffrey Ullman. Principles of optimal page replacement. *Journal of the ACM*, Vol. 18, Iss. 1, 1971. DOI: 10.1145/321623.321632. 59

[5] AMD. Revision guide for AMD family 10h processors. `http://developer.amd.com/wordpress/media/2012/10/41322.pdf`, August 2011. 3

[6] AMD. AMD64 architecture programmer's manual, rev. 3.24. `http://developer.amd.com/resources/documentation-articles/developer-guides-manuals`, October 2013. 36

[7] Nadav Amit. Optimizing the TLB shootdown algorithm with page access tracking. *USENIX Annual Technical Conference*, 2017. 44, 130

[8] Andrea Arcangeli. Transparent hugepage support. *KVM Forum*, 2010. 129

[9] ARM. *ARM Architecture Reference Manual*, 2013. 36

[10] Thomas Barr, Alan Cox, and Scott Rixner. Translation caching: Skip, don't walk (the page table). *International Symposium on Computer Architecture*, 2010. DOI: 10.1145/1815961.1815970. 3, 21, 36, 47, 98, 136

[11] Thomas Barr, Alan Cox, and Scott Rixner. SpecTLB: A mechanism for speculative address translation. *International Symposium on Computer Architecture*, 2011. DOI: 10.1145/2000064.2000101. 36, 97, 120, 121

[12] Arkaprava Basu, Jayneel Gandhi, Jichuan Chang, Mark Hill, and Michael Swift. Efficient virtual memory for big memory servers. *International Symposium on Computer Architecture*, 2013. DOI: 10.1145/2508148.2485943. 3, 21, 46, 136, 138, 139, 140, 141

[13] Arkaprava Basu, Jayneel Gandhi, Mark Hill, and Michael Swift. Reducing memory reference energy with opportunistic virtual caching. *International Symposium on Computer Architecture*, 2012. DOI: 10.1109/isca.2012.6237026. 28, 38, 39

[14] Lazlo Belady, Randolph Nelson, and Gerald Shedler. An anomaly in space-time characteristics of certain programs running in a paging machine. *Communications of the ACM*, 1969. DOI: 10.1145/363011.363155. 57

[15] Emergy Berger, Kathryn McKinley, Robert Blumofe, and Paul Wilson. Hoard: A scalable memory allocator for multithreaded programs. *International Conference on Architectural Support for Programming Languages and Operating Systems*, 2000. DOI: 10.1145/378995.379232. 61, 62, 65

[16] Emery Berger, Benjamin Zorn, and Kathryn McKinley. Reconsidering custom memory allocation. *Object-Oriented Programming, Systems, Languages and Applications*, 2002. DOI: 10.1145/582419.582421. 61

[17] Ravi Bhargava, Benjamin Serebrin, Francisco Spadini, and Srilatha Manne. Accelerating two-dimensional page walks for virtualized systems. *International Conference on Architectural Support for Programming Languages and Operating Systems*, 2008. DOI: 10.1145/1346281.1346286. 3, 47, 97, 98, 136

[18] Abhishek Bhattacharjee. Large-reach memory management unit caches. *International Symposium on Microarchitecture*, 2013. DOI: 10.1145/2540708.2540741. 3, 36, 47, 48, 97, 98, 102, 136

[19] Abhishek Bhattacharjee. Translation-triggered prefetching. *International Conference on Architectural Support for Programming Languages and Operating Systems*, 2017. DOI: 10.1145/3037697.3037705. 3, 21, 27, 36, 43, 122, 123, 124, 125, 126

[20] Abhishek Bhattacharjee, Daniel Lustig, and Margaret Martonosi. Shared last-level TLBs for chip multiprocessors. *17th International Symposium on High Performance Computer Architecture (HPCA)*, 2011. DOI: 10.1109/hpca.2011.5749717. 3, 21, 37, 41, 45, 46, 47, 101, 102, 103, 104, 130

[21] Abhishek Bhattacharjee and Margaret Martonosi. Characterizing the TLB behavior of emerging parallel workloads on chip multiprocessors. *12th International Conference on Parallel Architectures and Compilation Techniques (PACT)*, 2009. DOI: 10.1109/pact.2009.26.

[22] Abhishek Bhattacharjee and Margaret Martonosi. Inter-core cooperative TLB prefetchers for chip multiprocessors. *International Conference on Architectural Support for Programming Languages and Operating Systems*, 2010. DOI: 10.1145/1735971.1736060. 3, 45, 47, 102, 104, 130

[23] Jeff Bonwick. The slab allocator: An object-caching kernel memory allocator. *USENIX Annual Technical Conference*, 1994. 65, 66

[24] James Bornholt, Antoine Kaufmann, Jialin Li, Arvind Krishnamurthy, Emina Torlak, and Xi Wang. Specifying and checking file system crash-consistency models. *21st International Conference on Architectural Support for Programming Languages and Operating Systems (ASPLOS)*, 2016. DOI: 10.1145/2872362.2872406. 84

[25] Jacob Bramley. Page colouring on ARMv6 (and a bit on ARMv7). `https://community.arm.com/processors/b/blog/posts/page-colouring-on-armv6-and-a-bit-on-armv7`, 2013. 40

[26] Intel. Intel Broadwell specs. `http://www.7-cpu.com/cpu/Broadwell.html` 114

[27] Richar Carr and John Hennessy. WSCLOCK—a simple and effective algorithm for virtual memory management. *International Symposium on Operating Systems Principles*, 1981. DOI: 10.1145/800216.806596. 59

[28] Michel Cekleov, Michel Dubois, Jin-Chin Wang, and Faye Briggs. Virtual-address caches. *USC Technical Report*, No. CENG 09–18, 1990. 39

[29] Xiaotao Chang, Hubertus Franke, Yi Ge, Tao Liu, Kun Wang, Jimi Xenidis, Fei Chen, and Yu Zhang. Improving virtualization in the presence of software managed translation lookaside buffers. *International Conference on Computer Design*, 2001. DOI: 10.1145/2485922.2485933. 45

[30] Austin Clements, Frans Kaashoek, and Nickolai Zeldovich. Scalable address spaces using RCU balanced trees. *International Conference on Architectural Support for Programming Languages and Operating Systems*, 2012. DOI: 10.1145/2150976.2150998. 51

[31] Austin Clements, Frans Kaashoek, and Nickolai Zeldovich. RadixVM: Scalable address spaces for multithreaded applications. *European Conference on Computer Systems*, 2013. DOI: 10.1145/2465351.2465373. 51

[32] Guilherme Cox and Abhishek Bhattacharjee. Efficient address translation with multiple page sizes. *International Conference on Architectural Support for Programming Languages and Operating Systems*, 2017. 3, 21, 26, 42, 51, 114, 117, 118, 119

[33] Guilherme Cox, Zi Yan, Abhishek Bhattacharjee, and Vinod Ganapathy. A 3D-stacked architecture for secure memory acquisition. *Rutgers Technical Report DCS-TR-724*, 2016. 44

[34] Peter Denning. The working set model for program behavior. *International Symposium on Operating Systems Principles*, 1967. DOI: 10.1145/800001.811670. 56

[35] Peter Denning. Virtual memory. *Computing Surveys*, Vol. 2, No. 3, 1970. DOI: 10.1145/234313.234403. 3, 58, 59

[36] Peter Denning and Stuart Schwartz. Properties of the working-set model. *International Symposium on Operating Systems Principles*, 1972. DOI: 10.1145/800212.806511. 56

[37] Hugh Dickins. RMAP 17 real priotree. `https://lwn.net/Articles/82373/`, 2004. 66

[38] Xiaowan Dong, Sandhya Dwarkadas, and Alan Cox. Shared address translation revisited. *European Conference on Computer Systems*, 2016. DOI: 10.1145/2901318.2901327. 44

[39] Richard Draves. Page replacement and reference bit emulation in mach. *USENIX Mach Symposium*, 1991. 59

[40] Yu Du, Miao Zhu, Bruce Childers, Daniel Mosse, and Rami Melhem. Supporting super-pages in non-contiguous physical memory. *International Symposium on High Performance Computer Architecture*, 2015. DOI: 10.1109/hpca.2015.7056035. 134, 135, 136, 137

[41] Jake Edge. Kernel address space layout randomization. `https://lwn.net/Articles/569635/`, 2013. 56

[42] Dai Edwards. Designing and building Atlas. *Resurrection: The Bulletin of the Computer Conservation Society*, 62:9–18, 2013. 22

[43] Jayneel Gandhi, Arkaprava Basu, Mark Hill, and Michael Swift. Efficient memory virtualization: Reducing dimensionality of nested page walks. *International Symposium on Microarchitecture*, 2014. DOI: 10.1109/micro.2014.37. 21, 42, 46, 96, 141

[44] Jayneel Gandhi, Mark Hill, and Michael Swift. Agile paging: Exceeding the best of nested and shadow paging. *International Symposium on Computer Architecture*, 2016. DOI: 10.1109/isca.2016.67. 3, 96, 97, 98

[45] Fabien Gaud, Baptiste Lepers, Jeremie Decouchant, Justin Funston, Alexandra Fedorova, and Vivien Quema. Large pages may be harmful on NUMA systems. *USENIX Annual Technical Conference*, 2014. 129, 143

[46] Kourosh Gharachorloo, Daniel Lenoski, James Laudon, Phillip Gibbons, Anoop Gupta, and John Hennessy. Memory consistency and event ordering in scalable shared-memory multiprocessors. *17th International Symposium on Computer Architecture (ISCA)*, 1990. DOI: 10.1145/325164.325102. 88

[47] Cristiano Giuffrida, Anton Kuijsten, and Andrew Tanenbaum. Enhanced operating system security through efficient and fine-grained address space randomization. *USENIX Security Conference*, 2012. 55

[48] Jérôme Glisse et al. Heterogeneous memory management. `https://cgit.freedeskt op.org/~glisse/linux/log/?h=hmm-v25-4.9`, 2017. 87

[49] James R. Goodman. Cache consistency and sequential consistency. *Computer Science Department of Technical Report 1006*, University of Wisconsin-Madison, 1991. 39

[50] Leo J. Guibas and Robert Sedgewick. A dichromatic framework for balanced trees. In *19th IEEE Annual Symposium on Foundations of Computer Science*, pages 8–21, 1978. DOI: 10.1109/sfcs.1978.3. 52

[51] Haswell. Intel Haswell specs. `http://www.7-cpu.com/cpu/Haswell.html` 114

[52] HSA Foundation. HSA programmer's reference manual: HSAIL virtual ISA and programming model, compiler writer, and object format (BRIG), 2015. 88

[53] Jerry Huck and Jim Hays. Architectural support for translation table management in large address space machines. *International Symposium on Computer Architecture*, 1993. DOI: 10.1109/isca.1993.698544. 35

[54] IBM. Power ISA version 2.07, 2013. 34

[55] Intel. Intel 64 and IA-32 architectures software developer's manual. *Order Number 325462-048US*, September 2013. 24, 36

[56] Intel. 5-level paging and 5-level EPT. *Intel Whitepaper*, 2016. 10, 25, 26

[57] Bruce Jacob and Trevor Mudge. A look at several memory-management units, TLB-refill mechanisms, and page table organizations. *International Conference on Architectural Support for Programming Languages and Operating Systems*, 1998. DOI: 10.1145/291069.291065. 33, 45, 47

[58] Bruce Jacob and Trevor Mudge. Virtual memory in contemporary microprocessors. *IEEE Micro*, Vol. 18, Iss. 4, 1998. DOI: 10.1109/40.710872. 35

[59] Aamer Jaleel and Bruce Jacob. In-line interrupt handling for software-managed TLBs. *International Conference on Computer Design*, 2001. DOI: 10.1109/iccd.2001.955004. 45

[60] Jeongjin Jang, Sangho Lee, and Taesoo Kim. Breaking kernel address space layout randomization with intel TSX. *Conference on Computer and Communications Security*, 2016. DOI: 10.1145/2976749.2978321. 56

[61] Song Jiang, Feng Chen, and Xiaodong Zhang. CLOCK—Pro: An effective improvement of the CLOCK replacement. *USENIX Technical Conference*, 2005. 43

[62] Song Jiang, Xiaoning Ding, Feng Chen, Enhua Tan, and Xiaodong Zhang. DULO: An effective buffer cache management scheme to exploit both temporal and spatial locality. *USENIX Conference on File and Storage Technologies*, 2005. 61

[63] Song Jiang and Xiaodong Zhang. LIRS: An efficient low inter-reference recency set replacement policy to improve buffer cache performance. *International Conference on Measurement and Modeling of Computer Systems*, 2002. DOI: 10.1145/511334.511340.

[64] Stephen Jones, Andrea Arpaci-Dusseau, and Remzi Arpaci-Dusseau. Geiger: Monitoring the buffer cache in a virtual machine environment. *International Conference on Architectural Support for Programming Languages and Operating Systems*, 2006. DOI: 10.1145/1168857.1168861. 61

[65] Gokul Kandiraju and Anand Sivasubramaniam. Going the distance for TLB prefetching: An application-driven study. *International Symposium on Computer Architecture (ISCA)*, 2002. DOI: 10.1109/isca.2002.1003578. 104, 130

[66] Vasileios Karakostas, Jayneel Gandhi, Furkan Ayar, Adrian Cristal, Mark Hill, Kathryn McKinley, Mario Nemirovsky, Michael Swift, and Osman Unsal. Redundant memory mappings for fast access to large memories. *International Conference on Computer Architecture*, 2015. DOI: 10.1145/2749469.2749471. 3, 21, 141

[67] Vasileios Karakostas, Jayneel Gandhi, Adrian Cristal, Mark Hill, Kathryn McKinley, Mario Nemirovsky, Michael Swift, and Osman Unsal. Energy-efficient address translation. *International Symposium on High Performance Computer Architecture*, 2016. DOI: 10.1109/hpca.2016.7446100. 3

[68] Stefanos Kaxiras and Alberto Ros. A new perspective for efficient virtual-cache coherence. *International Symposium on Computer Architecture*, 2013. DOI: 10.1145/2508148.2485968. 39

[69] Richard E. Kessler and Mark D. Hill. Page placement algorithms for large real-indexed caches. *ACM Transactions on Computer Systems (TOCS)*, 10(4):338–359, 1992. DOI: 10.1145/138873.138876. 66

[70] Khronos Group. OpenCL 2.0. http://www.khronos.org/opencl 88

[71] Chongkyung Kil, Jinsuk Jun, Cristopher Bookholt, Jun Xu, and Peng Ning. Address space layout permutation (ASLP): Towards fine-grained randomization of commodity software. *Annual Computer Security Applications Conference*, 2006. DOI: 10.1109/ac-sac.2006.9. 55

[72] Donald Knuth. Fundamental algorithms. *The Art of Computer Programming*, 1997. 63, 65

[73] Bradley Kuszmaul. Supermalloc: A super fast multithreaded malloc for 64-bit machines. *International Symposium on Memory Management*, 2015. DOI: 10.1145/2754169.2754178. 62, 65

[74] Youngjin Kwon, Hangchen Yu, Simon Peter, Cristopher Rossbach, and Emmett Witchel. Coordinated and efficient hugepage management with INGENS. *International Symposium on Operating Systems Design and Implementation*, 2016. 21, 26, 51, 129

[75] Donghee Lee, Jongmoo Choi, Jong-Hun Kim, Sam Noh, Sang Lyul Min, Yookun Cho, and Chong Sang Kim. On the existence of a spectrum of policies that subsumes the least recently used (LRU) and least frequently used (LFU) policies. *International Conference on Measurement and Modeling of Computer Systems*, 1999. DOI: 10.1145/301453.301487. 43

[76] Daniel Lenoski, James Laudon, Kourosh Gharachorloo, W.-D. Weber, Anoop Gupta, John Hennessy, Mark Horowitz, and Monica S. Lam. The Stanford DASH multiprocessor. *Computer*, 25(3):63–79, 1992. DOI: 10.1109/2.121510. 88

[77] Linus Torvalds. Dirty Cow vulnerability in linux. https://lkml.org/lkml/2016/10/19/860 84

[78] Andy Lutorminski. Linux page table management memory ordering bug. https://lkml.org/lkml/2016/1/8/912 83

[79] Daniel Lustig, Abhishek Bhattacharjee, and Margaret Martonosi. TLB improvements for chip multiprocessors: Inter-core cooperative prefetchers and shared last-level TLBs. *ACM Transactions on Architecture and Code Optimization (TACO)*, April 10, 2013. DOI: 10.1145/2445572.2445574. 3, 101, 103, 104

[80] Daniel Lustig, Geet Sethi, Margaret Martonosi, and Abhishek Bhattacharjee. COATCheck: Verifying memory ordering at the hardware-OS interface. *21st International Conference on Architectural Support for Programming Languages and Operating Systems (ASPLOS)*, 2016. DOI: 10.1145/2872362.2872399. 3, 43, 84

[81] David Nagle, Richard Uhlig, Tim Stanley, Stuart Sechrest, Trevor Mudge, and Richard Brown. Design tradeoffs for software-managed TLBs. *International Symposium on Computer Architecture*, 1993. DOI: 10.1109/isca.1993.698543. 41, 45

[82] Juan Navarro, Sitaram Iyer, Peter Druschel, and Alan Cox. Practical, transparent operating system support for superpages. *International Symposium on Computer Architecture*, 2013. DOI: 10.1145/1060289.1060299. 22, 26

[83] NVIDIA. PTX ISA, Memory Consistency Model. https://developer.nvidia.com/cuda-toolkit 88

[84] Lea Olson, Jason Power, Mark Hill, and David Wood. Border control: Sandboxing accelerators. *International Symposium on Microarchitecture*, 2015. DOI: 10.1145/2830772.2830819. 93

[85] Mark Oskin and Gabriel Loh. A SW-managed approach to die-stacked DRAM. *International Conference on Parallel Architectures and Compilation Techniques*, 2015. DOI: 10.1109/pact.2015.30. 130

[86] Myrto Papadopoulou, Xin Tong, Andre Seznec, and Andreas Moshovos. Prediction-based superpage-friendly TLB designs. *International Symposium on High Performance Computer Architecture*, 2014. DOI: 10.1109/hpca.2015.7056034. 21, 42, 114, 115, 116

[87] Steven Pelley, Peter M. Chen, and Thomas F. Wenisch. Memory persistency. *41st International Symposium on Computer Architecture (ISCA)*, 2014. DOI: 10.1109/isca.2014.6853222. 84, 91

[88] Binh Pham, Abhishek Bhattacharjee, Yasuko Eckert, and Gabriel Loh. Increasing TLB reach by exploiting clustering in page translations. *International Symposium on High Performance Computer Architecture*, 2014. DOI: 10.1109/hpca.2014.6835964. 3, 28, 53, 107, 112

[89] Binh Pham, Derek Hower, Abhishek Bhattacharjee, and Trey Cain. TLB shootdown mitigation for low-power many-core servers with L1 virtual caches. *IEEE Computer Architecture Letters*, 2017. DOI: 10.1109/lca.2017.2712140. 44, 130

[90] Binh Pham, Viswanathan Vaidyanathan, Aamer Jaleel, and Abhishek Bhattacharjee. CoLT: Coalesced large-reach TLBs. *International Symposium on Microarchitecture*, 2012. DOI: 10.1109/micro.2012.32. 3, 21, 28, 51, 53, 107, 108, 109, 110, 111, 112, 113

[91] Binh Pham, Jan Vesely, Gabriel Loh, and Abhishek Bhattacharjee. Large pages and lightweight memory management in virtualized systems: Can you have it both ways? *International Symposium on Microarchitecture*, 2015. DOI: 10.1145/2830772.2830773. 3, 21, 22, 51, 55, 96, 97, 120, 121, 122, 129

[92] D. Pham, S. Asano, M. Bolliger, M. N. Day, H. P. Hofstee, C. Johns, J. Kahle, A. Kameyama, J. Keaty, Y. Masubuchi, M. Riley, D. Shippy, D. Stasiak, M. Suzuoki, M. Wang, J. Warnock, S. Weitzel, D. Wendel, T. Yamazaki, and K. Yazawa.

The design and implementation of a first-generation CELL processor, 2005. DOI: 10.1109/isscc.2005.1493930. 92

[93] Qualcomm. Qualcomm Snapdragon 810 processor, 2015. 86

[94] Bogdan F. Romanescu, Alvin R. Lebeck, and Daniel J. Sorin. Specifying and dynamically verifying address translation-aware memory consistency. *20th International Conference on Architectural Support for Programming Languages and Operating Systems (ASPLOS)*, 2010. DOI: 10.1145/1736020.1736057. 3, 43, 82

[95] Bogdan F. Romanescu, Alvin R. Lebeck, Daniel J. Sorin, and Anne Bracy. UNified instruction/translation/data (UNITD) coherence: One protocol to rule them all. *16th International Symposium on High-performance Computer Architecture (HPCA)*, 2010. DOI: 10.1109/hpca.2010.5416643. 44, 71, 127, 130

[96] Jee Ho Ryoo, Nagendra Gulur, Shuang Song, and Lizy John. Rethinking TLB designs in virtualized environments: A very large part-of-memory TLB. *International Symposium on Microarchitecture*, 2017. DOI: 10.1145/3079856.3080210. 105, 106, 107

[97] Ashley Saulsbury, Fredrik Dahlgren, and Per Stenstrom. Recency-based TLB preloading. *International Symposium on Computer Architecture (ISCA)*, 2002. DOI: 10.1145/339647.339666. 104, 131, 132, 133, 134

[98] Vivek Seshadri, Gennady Pekhimenko, Olatunji Ruwase, Onur Mutlu, Phillip Gibbons, Michael Kozuch, Todd Mowry, and Trishul Chilimbi. Page overlays: An enhanced virtual memory framework. *International Symposium on Computer Architecture*, 2015. DOI: 10.1145/2749469.2750379. 55

[99] Andre Seznec. Concurrent support of multiple page sizes on a skewed associative TLB. *IEEE Transactions on Computers*, 2003. DOI: 10.1109/tc.2004.21. 114, 115

[100] Hovav Shacham, Matthew Page, Ben Pfaff, Eu-Jin Goh, Nagendra Modadugu, and Dan Boneh. On the effectiveness of address-space randomization. *Conference on Computer and Communications Security*, 2004. DOI: 10.1145/1030083.1030124. 55

[101] Daniel Sorin, Mark Hill, and David Wood. *A Primer on Memory Consistency and Cache Coherence*. Synthesis Lectures on Computer Architecture. Morgan & Claypool Publishers, 2011. DOI: 10.2200/s00346ed1v01y201104cac016. 70, 82

[102] Dmitri B. Strukov, Gregory S. Snider, Duncan R. Stewart, and R. Stanley Williams. The missing memristor found. *Nature*, 453, May 2008. DOI: 10.1038/nature08166. 90

[103] Madhusudan Talluri and Mark Hill. Surpassing the TLB performance of superpages with less operating system support. *International Conference on Architectural Support for Programming Languages and Operating Systems*, 1994. DOI: 10.1145/195473.195531. 107

154 BIBLIOGRAPHY

[104] Madhusudan Talluri, Mark Hill, and Yousef Khalidi. A new page table for 64-bit address spaces. DOI: 10.1145/224057.224071. 33

[105] George Taylor, Peter Davies, and Michael Farmwald. The TLB slice—a low-cost high-speed address translation mechanism. *International Symposium on Computer Architecture*, 1990. DOI: 10.1109/isca.1990.134546. 39

[106] Rollins Turner and Henry Levy. Segmented FIFO page replacement. *Segmented FIFO Page Replacement*, 1981. DOI: 10.1145/1010629.805473. 57

[107] Unified Extensible Firmware Interface (UEFI) Forum. Advanced configuration and power interface specification, version 6.2. http://www.uefi.org/sites/default/files/resources/ACPI_6_2.pdf 91

[108] Girish Venkatasubramanian, Renato Figueiredo, Ramesh Illikal, and Donald Newell. A simulation analysis of shared TLBs with tag based partitioning in multicore virtualized environments. *Workshop on Managed Multi-core Systems*, 2009. 41

[109] Jan Vesely, Arkaprava Basu, Mark Oskin, Gabriel Loh, and Abhishek Bhattacharjee. Observations and opportunities in architecting shared virtual memory for heterogeneous systems. *International Symposium on Performance Analysis of Systems and Software*, 2016. DOI: 10.1109/ispass.2016.7482091. 93

[110] Carlos Villavieja, Vasileios Karakostas, Lluis Vilanova, Yoav Etsion, Alex Ramirez, Avi Mendelson, Nacho Navarro, Adrian Cristal, and Osman S. Unsal. DiDi: Mitigating the performance impact of TLB shootdowns using a shared TLB directory. *20th International Conference on Parallel Architectures and Compilation Techniques (PACT)*, 2011. DOI: 10.1109/pact.2011.65. 44, 78, 127, 130

[111] Matthias Waldhauer. New AMD Zen core details emerged. http://dresdenboy.blogspot.com/2016/02/new-amd-zen-core-details-emerged.html, 2016. 46, 51

[112] Emmett Witchel, Josh Cates, and Krste Asanović. *Mondrian Memory Protection*, Vol. 30, 2002. DOI: 10.1145/635506.605429. 2

[113] Zi Yan, Jan Vesely, Guilherme Cox, and Abhishek Bhattacharjee. Hardware translation coherence for virtualized systems. *International Symposium on Computer Architecture*, 2017. DOI: 10.1145/3079856.3080211. 44, 127, 130

[114] Ting Yang, Emery Berger, Scott Kaplan, and Elliot Moss. CRAMM: Virtual memory support for garbage collected applications. *International Symposium on Operating Systems Design and Implementation*, 2006. 61

[115] Idan Yaniv and Dan Tsafrir. Hash, don't cache (the page table). *International Conference on Measurement and Modeling of Computer Systems*, 2016. DOI: 10.1145/2896377.2901456. 33, 34, 47

[116] Ying Ye, Richard West, Zhuoqun Cheng, and Ye Li. Coloris: A dynamic cache partitioning system using page coloring. In *Proc. of the 23rd International Conference on Parallel Architectures and Compilation*, pages 381–392. ACM, 2014. DOI: 10.1145/2628071.2628104. 66

[117] Hongil Yoon and Guri Sohi. Revisiting virtual L1 caches: A practical design using dynamic synonym remapping. *International Symposium on High Performance Computer Architecture*, 2016. DOI: 10.1109/hpca.2016.7446066. 39

[118] Yuanyuan Zhou, James Philbin, and Kai Li. The multi-queue replacement algorithm for second level buffer caches. *USENIX Annual Technical Conference*, 2001. 61

Authors' Biographies

ABHISHEK BHATTACHARJEE

Abhishek Bhattacharjee is an Associate Professor of Computer Science at Rutgers University. His research interests are in computer systems, particularly at the interface of hardware and software. More recently, he has also been working on designing chips for brain-machine implants and systems for large-scale brain modeling. Abhishek received his Ph.D. from Princeton University in 2010. Contact him at `abhib@cs.rutgers.edu`.

DANIEL LUSTIG

Daniel Lustig is a Senior Research Scientist at NVIDIA. Dan's work generally focuses on memory system architectures, and his particular research interests lie in memory consistency models, cache coherence protocols, virtual memory, and formal verification of all of the above. Dan received his Ph.D. in Electrical Engineering from Princeton in 2015. He can be reached at `dlustig@nvidia.com`.

Printed in the United States
by Baker & Taylor Publisher Services